# HOW TO SAVE OUR
# PLANET

# HOW TO SAVE OUR PLANET

## E. ROBERTS ALLEY

XULON PRESS

Xulon Press
2301 Lucien Way #415
Maitland, FL 32751
407.339.4217
www.xulonpress.com

© 2020 by E. Roberts Alley

All rights reserved solely by the author. The author guarantees all contents are original and do not infringe upon the legal rights of any other person or work. No part of this book may be reproduced in any form without the permission of the author. The views expressed in this book are not necessarily those of the publisher.

The cover of this book is an oil painting of Bent Lake, Ontario, Canada, by the author, expressing my goal of coexistence with nature in order to provide survival and sustainable growth for our planet.

Unless otherwise indicated, Scripture quotations taken from the English Standard Version (ESV). Copyright © 2001 by Crossway, a publishing ministry of Good News Publishers. Used by permission. All rights reserved.

Scripture quotations taken from the New American Standard Bible (NASB). Copyright © 1960, 1962, 1963, 1968, 1971, 1972, 1973, 1975, 1977, 1995 by The Lockman Foundation. Used by permission. All rights reserved.

Printed in the United States of America.

Paperback ISBN-13: 978-1-6312-9350-4

Ebook ISBN-13: 978-1-6312-9351-1

# Dedication

This book is dedicated to Jesus Christ and the other very important people of my life: my late wife, Marion; our children Rob, Lea, Aime, and Emma; their spouses: Fronda, Steve, Scott and Jeff; and their children: Reagan, Hayes, Gracey, and Otis; Will and Cooper; Catherine and Wesley; and Jimmy, Mimi, and Lily. I love you all!

# Table of Contents

Dedication................................................. v
Foreword............................................... xiii
    Are our Minds Capable of Saving our Planet? ................... xiii
    Will Saving the Environment Save our Planet?................... xiii
    Have we Started a Culture War which is Slowing us down? ........ xiii
    Comparing the Sides of the Culture War........................xvii
    Big Versus Small Government................................xvii
    Redistribution of Wealth .................................... xviii
    Health Care .............................................. xviii
    The Right to Life .......................................... xviii
    Equal Rights for All......................................... xix
    Protection of the Environment................................. xxi
    The Potential Culture War Resolution........................... xxi
Preface ................................................. xxv
**BOOK I: THE QUESTION** .............................. **XXIX**
The Need to Answer the Question ............................. 1
    The Purpose of This Book..................................... 1
    Does Our Planet Need Saving?................................. 3
    Evidence of Climate Change or Global Warming................... 4
    Hurricanes Versus Carbon Emissions ........................... 7
    Weather and Climate Disaster Costs Versus
    Carbon and CO2 Emissions.................................... 9

Does Human Activity Cause Melting
Ice Caps and Rising Sea Levels?...............................10
Solar Radiation Versus Global Temperatures ....................10
Global Temperatures Versus Carbon Emissions ..................11
Figure 1-4 ................................................11
Figure 1-5 ................................................12
Figure 1-6 ................................................13
CO2 Versus Altitude .......................................14
Figure 1-7 ................................................14
Temperature Distribution by Continent........................14
Summary of Record High and Low Temperatures in the U.S.A. ......15
Conclusion of the Evidence of Climate Change or Global Warming ...16
Is There Both a Natural and a Spiritual World?...................21
The Natural World as Experienced by Our Senses ................23
The Spiritual World ........................................26
How Does the Bible Fit In?...................................27
Using Our Humanity to Think................................28
Thinking About the Environment ............................29

# BOOK II: WHAT IS THE SCIENTIFIC MIND? ............ 33

Existence And Reality........................................35
   Introduction ...............................................35
   What is Existence?..........................................35
   What is Reality?............................................38
   Natural Reality.............................................39
   Supernatural Reality .......................................39
   A History of our Effort to Define Existence and Reality.............39
   A Brief History of Neuroscience ..............................40
   A Brief History of Philosophy ................................43
   A Brief History of Theology .................................46
   What Do These Histories Tell Us?..............................50
   The Importance of Natural Reality ............................58
   The Natural Permanence of Non-Living Matter ..................62

    The Natural Death of Living Matter ........................... 63
    Do We Accept a Supernatural Reality? ......................... 64
    Is the Supernatural Random? ................................. 66
    Beyond our Senses ........................................... 67
What is Humanity? ............................................... 71
    Introduction ................................................ 71
    Our Limited Intellect ....................................... 72
    What is Thinking? .......................................... 74
    What is Critical Thinking? .................................. 77
    How to Think Spiritually ................................... 79
    Thinking with Imagination .................................. 80
Our Body .......................................................... 81
    Introduction ................................................ 81
    DNA, Our Genetic Nature .................................... 82
    Theoretical Genetics ........................................ 85
Our Brain ......................................................... 89
    Introduction ................................................ 89
    Left Brained/Right Brained ................................. 90
Our Mind .......................................................... 93
    Introduction ................................................ 93
    The Natural Mind ........................................... 93
    What is Our Mind Biblically? ............................... 94
    The Limits of the Mind ..................................... 96
Our Soul .......................................................... 99
    Introduction ................................................ 99
    The Biblical Soul ........................................... 99
Our Spirit ........................................................ 103
    Introduction ............................................... 103
    The Holy Spirit ............................................ 104

# BOOK III: WHAT IS OUR PLANET? ................... 107
The Original Quality of the Environment .................... 109
    Introduction ............................................... 109

    The Air Environment............................................111
    The Water Environment..........................................113
    The Land Environment...........................................114
    Our Relationship to the Environment.............................116

The Residuals Entering Our Environment................................117
    Introduction....................................................117
    Natural Pollutants..............................................117
    From Land.......................................................118
    Pollution by Agriculture........................................122
    Pollution by Municipalities.....................................122
    Pollution by Industries.........................................123
    Pollution by Personal Activities................................125
    Estimating Residual Discharges..................................129
    Calculations for Direct Discharges..............................129
    Calculations for Indirect Discharges............................130
    Estimating Transportation Discharges............................131
    Public Transportation...........................................131
    Other Discharges................................................132
    "Green" Solutions Which do Not Solve the Problem................132

The Regulation of Residuals and Environmental Quality.................135
    Introduction....................................................135
    A Rational Method for Pollution Control.........................136
    Air Management through Regulations..............................140
    Air Problems which Remain.......................................141
    Water Management through Regulations............................143
    Water Problems Which Remain.....................................144
    Land Management Through Regulations.............................145

**BOOK IV: OUR LIMITED NATURAL RESOURCES.........151**

The Preservation of Natural Resources.................................153
    Natural Resources used by Humans................................155
    Natural Resources Used for Food.................................155
    Natural Resources used for Clothing.............................157

Natural Resources used for Shelter ............................. 157
Our Choices for Food Residuals Disposal ....................... 163
Our Choices for Clothing Residuals Disposal ................... 164
Our Choices for Shelter Residuals Disposal .................... 164
Our Choices for Transportation Residuals Disposal ............. 165
Our Choices for Recreation Residuals Disposal ................. 166
Our Choices for Electronics Residuals Disposal ................ 166
Summary of Our Choices ........................................ 166

# BOOK V: HOW WE CAN SAVE OUR PLANET ............ 169

How We as a Society Can Save Our Planet .................... 171
    Introduction ............................................... 171
    The Preservation of Humanity ............................... 172
    The Biblical View of the Preservation of Humanity .......... 173
    The Secular Effort to Preserve Humanity .................... 174
    Stem Cell-Based Therapies .................................. 176
    Immunotherapy .............................................. 177
    Therapeutic Viruses ........................................ 178
    Cancer Tumor Starvation .................................... 178
    Microsurgeries ............................................. 178
    Nanotechnology ............................................. 178
    3D Printing ................................................ 178
    Summary of the Secular Effort to Extend Our Lives .......... 179
    Is There Natural Reality Beyond Our Present Knowledge? ..... 181
    Fractons ................................................... 183
    What is the Potential for Understanding these Microparticles? ...... 183
    How Can Our Societal Actions Save the Air? ................. 185
    Carbon Dioxide Control ..................................... 187
How We as Individuals Can Save Our Planet .................. 203
    Introduction ............................................... 203
    Residuals Accounting ....................................... 203
    The Mission Statement ...................................... 204
    Plan Boundaries ............................................ 205

    Geographical Boundary..................................205

    Organizational Boundary................................205

    Life Cycle Analysis......................................205

        A Summary of Practical Ways to Save
Our Environment as Individuals............................207

What is the Ultimate Fate of Our Planet?....................209

    Introduction............................................209

    The Natural Reality.....................................210

    Spiritual Reality........................................211

    Outer Space............................................212

    Observable Nature.....................................213

    Nanospace.............................................213

    If Not Outer Space, Nature and Nanospace,
What is the Answer to Our Fate?............................214

    If We Do Nothing, What Then?..........................214

    What is the Potential for Saving Our Planet?.............216

The Spiritual Future of our Planet..........................221

    Introduction............................................221

    Why Quote the Bible When We are
Studying the Mind and Environment?.......................221

    The Truth of the Bible..................................222

    The Kingdom of God and the Kingdom of Heaven..............236

## BOOK VI: A CHALLENGE TO SUCCEED................243

Conclusion: Our Natural Planet Can be Saved................245

    Introduction............................................245

    The Challenge..........................................245

    The Societal Pressure to Defeat our Challenge............246

    One Practical Possibility of Financing the Challenge as an Example..249

    A Caution..............................................251

Conclusion................................................253

About the Author..........................................255

Books by E. Roberts Alley..................................257

# Foreword

### Are our Minds Capable of Saving our Planet?

This book will first answer the basic question of whether our individual and corporate scientific minds are capable of developing policies and procedures which can save our planet. If not, this goal is hopeless from our perspective.

### Will Saving the Environment Save our Planet?

The second major question addressed in this book, is what is our environment, and is it past the time when it is savable. Then we will concentrate on the issues which remain unsolved, and suggest steps to immediately resolve these issues.

### Have we Started a Culture War which is Slowing us down?

An historian will look at the last decades of the 20$^{th}$ Century, and the first decades of the 21$^{st}$ century, as an odd Culture War between those who call themselves *Progressives* and their enemies who call themselves *Conservatives*. The *Progressives* tend toward socialism; big government, redistribution of wealth, reproductive rights, free education, medical care and housing for all – or at least the poor.

The *Conservatives* tend toward representative government, state's rights, free enterprise, the right to life, education, medical care, and housing by choice.

The vitriol in this war has reached levels of hatred where the sides describe their enemies as either Communists or Nazis.

The *Progressives* are using diversity, climate change, and rights of selected groups as arrows in their quiver to demonstrate the insensitivity of the *Conservatives* to those unable to enjoy equal standards of living. On the other hand, *Conservatives* are using job production, the right to succeed, and religious rights as their arrows to demonstrate the irrationality of the *Progressives* for not allowing the reward from the hard work of the populace striving to raise their standard of living.

According to published platforms and position statements, the faith of *Progressives* is vested in humans, as described in the *Humanist Manifesto I* by Raymond Bragg, New *Humanist*, May/June, 1933; *Humanist Manifesto II*, Paul Kurtz, Edwin H. Wilson, *The Humanist*, September/October, 1973; *Humanism and Its Aspirations*, *Humanist Manifesto III*, American Humanist Association, 2003. According to these manifestos, as essentially adopted by Socialist and Democratic platforms, *Progressives* have consistently assumed that humans, in the form of leadership and government, possess the ideas, the ideals, and the sense of justice to promote an egalitarian world community based on voluntary mutual cooperation; and oppose an acquisitive and profit-motivated society, as a new religion (Humanist *Manifesto I*).

The *Humanist Manifesto II* rejected theism and deism and opposed weapons of mass destruction. It also supported human rights, proposed an international court, and advocated for the right to unrestricted abortion and contraception.

*Humanist Manifesto III,* promotes empiricism, the belief that the knowledge of the world is derived solely by observation, experimentation, and rational analysis. It also promotes the belief that humans are derived from unguided evolutionary change; and ethical naturalism, as well as the belief that values are derived solely from human need and interest.

As a result of these beliefs, *progressives*, reject the belief in God as important, and promote the prohibition of its mention or explanation in public, especially the ever expanding governmental portion of the "public". As their god, they normally accept science as capable of determining truth, fairness and justice. They typically promote higher taxation on the wealthy, and redistribution of this federal income to the less wealthy, to assure an equal standard of living, regardless of risk, responsibility,inheritance, or education. As one means to this goal, *progressives* normally promote unions to force the redistribution of profits to employees. They believe in equal rights for all, regardless of race, religion, gender, perceived gender or social standing, a position shared with *conservatives*. But the *progressives* tend to go farther, even including reparations for past injustices incurred. There is an open invitation to add new Civil rights to those granted through the Civil rights Act of 1964. Since then, originally unnamed rights have been given to races, genders, including transgender, sexual orientation and immigrants, not by the Legislative, but the Executive or Judicial branches of the government. By listing these beliefs, my intention is not to say that they are invalid, especially the equal rights for all statement.

The exception to these personal rights is the right of the unborn to live. *Progressives* would allow them to be murdered for the convenience of their parents, other interested parties, or because they may not be what the *Progressives* deem normal, or of the proper gender.

On the other hand, the faith of *Conservatives* is typically invested in God, as described in the Old and New Testaments of the Holy Bible. They hold to the belief that the knowledge of the world is found in both the natural world and the spiritual world. They typically believe that life was created by God rather than having been evolved from an unknown source and that God intended us to be good and fair to all by obeying His commandments. Other than party platforms, books and articles, there has been less consistency as to specific beliefs from the *Conservatives*, probably because most Jews and Christians believe that the Bible interprets the Bible, and believers have the right to follow their own hearts.

In response to the expressed and consistent *Progressive* beliefs, *Conservatives* typically believe that leadership and government exist to serve the people, instead of vice versa. They believe that their government's ideas, ideals, and justice exist to provide the people with a fair opportunity to survive and advance their position in society. *Conservatives* typically believe in governmental fiscal accountability, such as a balanced budget, as representatives of the people. They accept redistribution of wealth as unfair toward the hard work of an individual or a family. They believe that property rights are individual rights. They believe in the ultimate fairness of a profit-motivated free enterprise society. They believe in family rights and values and the moral code expressed in the Bible by God, rather than by humans. They accept the *US Constitution* as the guide to our legal system, and state's rights as a provision of the *Constitution*. They prefer that laws be enacted, interpreted and enforced by their elected representatives, rather than the Executive or Judicial branch of the government.

They also reject weapons of mass destruction as used historically mostly by *Progressive* governments. *Conservatives* normally promote limited foreign involvement, aid, and court systems. They almost always consider every person to have equal rights, regardless of age, race, religion, or social; standing, and reject the right to murder unborn humans. They typically believe that the government should not intervene in sexual relations between adults and should not be involved in promoting such in children or adults. In summary, *Conservatives* typically believe that ethics, fairness, and justice are established and defined by God, as expressed in His Word, and humans are subservient to that truth.

## Comparing the Sides of the Culture War

Ideally we should follow a standard in order to discern which of the divergent beliefs in the Culture War has the highest probability of saving our planet. Simplifying, the divergent beliefs are:

- Big versus small government
- Redistribution of wealth
- Health care
- The right to life
- Equal rights for all
- Protection of the environment

Each of these basic differences will be analyzed below as to the effect they could have on the survival of our planet. My attempt in this comparison is to differentiate between *permitted* and *promoted* cultural controls in order to predict the ultimate outcome should virtually all people become affected.

## Big Versus Small Government

The type or size of the government we have does not seem to affect the survival of the planet. Historically, more democratic or representative governments have succeeded long term than socialistic, but the civilizations have continued in either case. Amendment 10 of the Constitution should resolve this issue if indeed we practice: *The powers not delegated to the United States by the Constitution, nor prohibited by it to the States, are reserved to the States respectively, or to the people.* The conflict seems to be more in terms of control, freedom and potential wealth accumulation than survival.

## Redistribution of Wealth

This conflict of ideas likewise should affect the technological and cultural advancement of a civilization but not whether the planet will survive. Theoretically though, the more poverty, the more premature deaths, which could cause a lack of growth of a society, but probably not a cessation. This issue should be solvable by fairly determining who is eligible to receive more wealth, who should be punished for having too much wealth, who is truly unable to work, and who just chooses not to work.

## Health Care

Both participants in the War believe in medical care available to all, The *progressives* usually believe that all, or at least a majority, of medical care should be required, and paid for by the federal government. The *conservatives* seem to believe that medical care should be provided by employers, and for those without jobs, provided by the federal government. The problem caused by the *conservatives* is that they will not require all employers to provide medical care, regardless of hours worked. The problem caused by *progressives* is the differentiation between can't get a job, and won't get a job. The wrong solution of this issue could cause unnecessary deaths, potentially leading to the collapse of society, if not resolved. A good example is the Corona Pandemic of 2020.

## The Right to Life

This is one of the few issues in the Culture War which could affect the survival of the planet. If we assume, as is possible, that the trend of abortion extends until all babies are murdered or sacrificed, then there would be no survival of a next generation. If a compromise were to be reached, as is probable, where only certain babies would be killed, the results could vary from a stagnant population, which some would prefer, to a radical population decline which could lead to the failure of civilization. As discussed in Chapter 13, the more population we have, the more brains we have, and the more resources to develop policies and practices to save our planet. As discussed below, it is unreasonable and arrogant to assume that only the educated and the powerful have answers to saving the planet.

## Equal Rights for All

In the Culture War, for some reason, there is no interest in equal rights for all by either side, but only equal rights for certain selected groups pf people. Exclusive rights, acceptance and even kindness to one group only, is not justice.

The U.S. Constitution provides for freedoms of religion, speech, the press, assembly and redress of grievances (Amendment 1), other rights retained by the people (Amendment 9),freedom from slavery (Amendment 13), equal rights of citizenship (Amendment 14), rights of people of color, race and previous servitude (Amendment 15), the right for both sexes to vote (Amendment 19, but no other equal rights for expressed groups of people. So for one of the government branches to grant rights to other groups of people, requires further action than the Constitution provides. These further rights could be for the wealthy, the rulers, the poor, the disabled or minorities. The groups most discussed in the Culture War are the last three categories. The first two are more related to the government size and the wealth distribution discussed above, and are more an issue of fairness, justice, health and safety than the question of equal rights for all.

Under discussion are several types of minorities including ethnic, religious, gender and sexual orientation. Of these, ethnic, religious and gender rights are provided for in the Constitution, and mustn't be denied. Since there are only two genders, there should be no controversy that both must be treated equally. It seems that every day more sexually related minorities are identified with a crucial need for equal rights. Why is sexual activity a minority category? It is practiced by a vast majority of the population. If each specific type of sexual activity were considered a minority qualification, the complications would be unending and far beyond the constitutional responsibilities of the federal government.

Considering sexual activities a governmental right, bizarre as it seems, is completely unnecessary. First and foremost, any person has the natural right to love their own sex equally or more than the other sex. The purpose of this book is not to debate rights, but to determine what practices could save our planet. Homosexuality can be defined as *agape,* genuine or godly love, or *phileo*, brotherly love. Two older men or women, or even college roommates would probably qualify for this category. They love or like those of the same sex and may live together, but not have sexual relations.

Neither side of the war normally speaks of these types of love. They are more interested in *eros,* sexual love. So what do the *progressives* want concerning equal rights for sexual love, and will the granting of those rights affect the permanence of human life on our planet? If we face the question of "what if everyone exclusively exercised homosexual intercourse", the obvious answer is that there would be no more births, and human life would cease. Using the *normal/natural* test, homosexual intercourse is not *normal*, in that only around 3.5% of the population (The Williams Institute, UCLA School of Law, April, 2011) practices homosexuality.

It is also not *natural,* since it is not prevalent in the animal kingdom. Heterosexual sex in plants and lower animals is practiced in the open with no guilt, shame or remorse. That tendency is part of their DNA and has the obvious purpose of procreation.

The old argument of gay heredity has been proven invalid in a recent test of over 500,000 people in the US and UK, by the Broad Institute of MIT and Harvard. Lead author Andrea Ganna, *Science*, August 30,2019, states that "There is no single continuum from opposite to same sexual preference" and "All tested genetic variants…do not allow meaningful prediction of an individual's sexual behavior."

The right of respect and acceptance for homosexuals, even though not specifically granted in the Constitution, is certainly expected in a civilized society.

So the trend of government promotion of this particular minority right of homosexual intercourse could indeed lead to the end of our planet since procreation is impossible as an ultimate result.

## Protection of the Environment

At the time the Constitution was written, the protection of the environment was not considered an important issue, therefore it is one of the few arguments of the Culture War that is not addressed to some extent in that document. The main purpose of this book is to help to bring this issue into consideration. *Progressives* tend to be more concerned with increasing the control of the federal government over its constituents, especially industries, than they are in saving the planet. *Conservatives* tend to value property rights in the form of the freedom to act in the interest of the property owner, even at the expense of the public. The purposes of the Constitution was to promote the general welfare, and secure the blessings of liberty, but an equal purpose was to insure domestic tranquility. It was not the purpose to consider these goals conflictatory, but complimentary. Therefore, it is our responsibility, to insist that acceptable freedoms and liberties be granted to citizens, without causing a lack of the same in others.

## The Potential Culture War Resolution

If the descriptions of the opposing platforms of the Culture War as expressed above are accurate, the only three controversies which could possibly destroy our planet, are abortion, a promotion of homosexuality and the destruction of our environment. These issues, if carried to their extreme, will cause a cessation of births as well as an end of nature, therefore an end to society. One caution in proceeding with the establishment of rights, privileges, rules and regulations that could resolve our Culture War, is that they should be fair to all citizens and not reserved for only one segment of society. For instance, if civil unions are allowed for homosexuals, they must be allowed for any two people co-habituating, regardless of gender of age, over an established minimum. That means, college roommates, two sisters or friends living together etc. Another example is that the right to use private property does not extend to polluting the air, water or soil, since private property always affects adjacent private property.

So logic would conclude that abortion is murder and if allowed to proliferate unchecked, will eventually lead to the death of humanity on our planet. But on the other hand, we should be able to resolve the requested homosexual civil rights issues, except for the desire to normalize, naturalize or promote homosexual intercourse. Homosexuals must be allowed to make their own decisions about their own lives, and the rest of society should respect those decisions. The only constraint we must have as a society is that we not allow the promotion of homosexual intercourse to the point that we will cease the advancement of procreation, and thereby, our civilization. Thirdly, we must protect all of the environment, even that which is privately owned from lethal damage.

It is sad to think that we may be fighting the Culture War over wealth, power, pleasure and satisfaction. It seems a waste of time, effort and money to disagree on whether we are satisfied with our wealth, whether others have more wealth, whether we don't want the burden of children, or desire to return to the adolescent pleasures and excitement of our past. Likewise, it is immature to allow one person or industry to pollute the nature that belongs to all of us.

**Typically left behind in this Culture War is the future of our planet and the environment of which it consists.** All of the issues occupying our emotions and efforts pale when we realize that with our present environmental practices, our planet will not survive. The *Progressives* blame our environmental crisis on human caused climate change, which this book will demonstrate as unproven.

The *Conservatives* believe that we can do anything to our environment without future repercussions, and that global warming is not occurring, both of which this book will also demonstrate are unsubstantiated. So, both sides of the Culture War are woefully incorrect in their efforts, and must get together to have any hope of saving our planet.

If we can extract ourselves from an unnecessary obsession with our emotional beliefs concerning the issues of the Culture War, we can instead, begin a journey of listening to, respecting, and working rationally and reasonably with each other. This will result in a process that can resolve a much larger question than those first five listed above. **Can we save our planet and enjoy everlasting peace, and if so, how?**

# Preface

As humans, we find ourselves trapped on this planet Earth. Interestingly, unlike the other living plants and animals, which are stuck here with us, we possess a mind capable of thinking, reasoning, and solving problems, in addition to simply existing. Most of our fellow humans, it seems, are satisfied to play the plant and animal game, by being born through no choice of their own, being trained through no choice of their own, at least initially. Then the humans exist for a few years on their own with no real impact on other humans, other than their families and friends. As such, those members of our society mostly are takers, rather than givers, users, rather than providers. They may have no interest in or effect on the survival, peace, or comfort of their fellow humans, or even the plants and animals under their care.

But, since we do possess this unique mind, should it be used to assure the survival of our civilization, or will we allow our species to become extinct through complacency, or through the distraction of the Culture War? Any thoughtful person can readily observe that our planet is rapidly deteriorating as a stable environment for human habitation.

Our initial problem in facing this reality is that we have no clue as to the length of time our planet will be able to support life.

The good news is that we have seen over a short period of time the exponential advancement of science, technology, engineering, and medicine. All are allowed by the efforts and results of our minds and the minds of those who proceeded us. But, even so, we seem to be quickly depleting our limited natural resources and destroying the environment of our planet in a way that is also exponentially moving forward.

This book will delve into the capability of our minds and the strength of our wills, to see if we can possibly stop or slow down this trend toward ultimate destruction. It will then examine the scientific characteristics of the environment in order to determine what we can practically accomplish.

**The timing of this effort to survive is up to us – both as a society and as individuals.**

Too many writers, broadcasters, and environmentalists have approached this issue emotionally instead of scientifically. By doing so, they have actually slowed our progress toward maintaining a habitable planet. I will attempt to restrict my comments in this book concerning the mind and the environment to documentable science. Any statement not documented should be accepted as only an opinion.

I have written this book as a scientist and a Christian, rather than a theologian. I am comfortable with my scientific observations and conclusions because of my education and experience. For the spiritual portions of this book, I rely on my lifelong Christian walk, my responsibility for about twenty years for Christian education in my church, my forty-five years of teaching Sunday school and Bible studies, but more than all of these, I rely and trust the Holy Spirit in me to interpret the Bible correctly. I hope that the spiritual statements presented will be a fresh look at the truth of the Bible, not encumbered by theology and seminary studies, which, face it, are taught by humans, not directly by God, as is the Holy Bible.

*Preface*

For some readers, this book may delve into tedious detail in subjects of no interest. If you find yourself in that position, I would suggest studying the Table of Contents in order to concentrate on the most appropriate sections for you. For those who read the entire book, there are some repetitions that exist because of this invitation to read only certain portions.

Through the process of the discovery of our capability for saving our planet, we will also look into the probability of the existence of a separate supernatural or spiritual reality, and its relationship to the scientifically discernable natural reality, and whether that relationship affects the future of our planet.

# BOOK I: THE QUESTION

*Is our Corporate Scientific Mind Capable of Saving the Planet?*

*Chapter 1*

# The Need to Answer the Question

## The Purpose of This Book

Is it possible to save our planet, or should we, as discussed in the Preface, just be content with living our brief lives, and leaving the future to the future? Can we call for peace in the Culture War and work together in a concerted effort toward a common goal?

In Book I, we answer this question posed in the title to Book I above, positively, and delve into the reality of our planet. Is it totally natural, even in its creation, or is it partly supernatural in its creation, its existence, and its fate? This question involves using our human capacity to *think* beyond mere animal existence. In this book we discuss data and graphs , mostly produced by the U.S. Government, which relate climate issues to human caused discharges into the air, and introduce the concept of both a natural and a spiritual reality.

In Book II, we present studies and theories of existence and reality, how we think, and how we have thought throughout history, about reality.

In order to try to understand the limitations of our minds, we will look at both the natural and the supernatural realities, how mankind throughout history has interpreted these realities, and how the emphasis has changed throughout history. We will determine whether the history of thinking about reality shows cyclical variability (history repeating itself), or a progressive or regressive movement. From this analysis, we will attempt to extrapolate the predictable future direction of the progress of our thinking; to see if indeed, our minds are capable of saving our planet.

We will then go into some detail about our thinking process and the tools of our body, brain, mind, soul, and spirit, which affect our thinking and our actions.

The purpose of Book III is to critically examine the history, the current state, and the potential of our environment, in order to determine whether our minds will provide us with the capability of either improving our environment. Or at least slow the deterioration of the air, water and land.

We will also discuss whether our three natural constituents, the *body, brain, and the mind,* are sufficient to save our planet; or whether the two supernatural constituents, the *soul and the spirit,* are required to accomplish this goal.

In Book III, we examine our environment; *not what it seems to be, or what we read that it is, but what science proves that it is,* and whether and how we can improve that environment.

Book IV concentrates on the often overlooked probability of us running out of critical limited natural resources, which are necessary for the manufacturing of products required for our growth and/or survival.

Book V looks at how we as a society, and how we as individuals can practically work together to save our planet. This Book also delves into the scientifically avoided area of the natural versus the supernatural states of reality and how they will affect our future. Book VI is a challenge to succeed in saving our natural planet, given the minds, wills, and sacrifices required to do something other than whine and blame.

When we compare these study tasks for this book, we must discern whether there is enough accumulated knowledge of the environment available, what other information is needed, and where that leads us in the future.

Please don't become discouraged because you are just one person. The personal percentage of the residuals of our civilization contributes around 87 percent of the capacity of landfills (see Chapter 11), and all municipalities, industries commercial establishments, etc. are managed by individuals like us. So we indeed can potentially control the future of our planet. Do we have the vision, will, and integrity to do it? We will explore in this book how this can be done on a corporate, governmental, and personal basis. After all, we affect, even control, all environmental destruction, and all-natural resource use; or we certainly should.

## Does Our Planet Need Saving?

Virtually every day, I read or hear of someone attempting to manipulate my emotions to the point that I will fear for the future of our planet. In the 1970s, the threat was *global cooling*. In the 1980s and 1990s, that threat was miraculously dropped in favor of *global warming*, the opposite prediction. Since 2000, the dire prediction has inexplicably been changed into *climate change*.

In all of these examples of contradictory reasoning, I have attempted to follow the logic to see if there is indeed a concern for the future of our planet. Part of my interest has been my profession for the past fifty-eight years in environmental science and engineering. During all of this experience, I have been a consulting scientist and/or engineer, author, teacher or professor, and have garnered knowledge and experience through worldwide involvement in the causes of environmental deterioration and the potential solutions to this concern. I must confess that my second motivation is that I was taught from an early age to think critically and to not accept statements without documented evidence. The mild cynicism associated with this tendency to think critically leads me to suspect that many of the emotional and alarmist statements of today are being made by those with a vested interest in the popular acceptance of the fear for survival.

For instance, weather forecasters are more popular, therefore, more powerful, during or prior to bad weather. The author of one of the alarmist books published during this latest period was an owner of a company that sold carbon credits. Consultants are only needed when a problem is perceived.

So when these contradictory beliefs began to be widely accepted, I began to research in the fields of my expertise to reach a conclusion which I could apply to my teaching and consulting. I tried to be logical in this research, which I based on deductive reasoning. I have had the honor of acting as an expert witness in several trials involving environmental issues and have had the opportunity to convince a judge and/or jury of the scientific truth. In every case, I found that if I restricted myself to scientific facts, and avoided the temptation to reach an undocumented conclusion, my explanations would prevail. I offer this brief background because that is the basis for the presentations and conclusions of this book. The book, as I intend it, is a summary of the answers to the statement posed in the title, *How to Save our Planet* and the presentation of the potential areas of science and history that could address the question.

## Evidence of Climate Change or Global Warming

### Natural Disasters Versus Carbon Emissions

In order to address the current concern for climate change, a summary of current objective data relating to weather as indicated by natural disasters can be found in the NOAA publication *Billion-Dollar Weather and Climate Disasters: : Overview*, through 2019, which lists all billion dollar plus storms, winter storms, droughts, floods, hurricanes and wildfire costs, CPI adjusted. This table is compared as a cost per ton to the Total U.S. Greenhouse Gas

Emissions, 1990-2017 as published by EPA, and shown on Figure 1-1 below.

1990-1999: $26.82 billion/yr vs. 6400-7000 million tons of carbon/year; 52 events (5.2/year) at an average of **$3.94/T (10 years)**

2000-2009: $50.68 billion/yr vs. 7200-6800 million tons of carbon/year; 59 events (5.9/year) at an average of **$7.24/T (10 years)**

2010-2017: $82.8 billion/yr vs. 6900-6500 million tons of carbon equivalent/year; 91 events (11.4/year) at an average of **$12.36/T (8 years)**

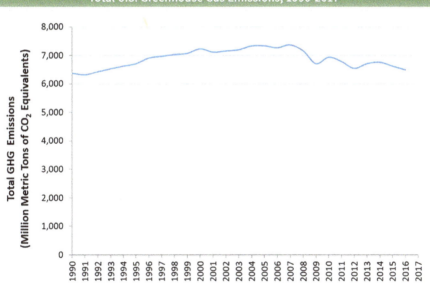

**Figure 1-1**

The conclusion is that there is no direct correlation between the costs of natural disasters and ghg emissions in the U.S.

From the 1980s to the 1990s, the number of natural disasters in the U.S. increased by 86% while the average ghg emissions increased by only 16%. From the 1990s to the 2000s, the number of natural disasters increased by 13% while the average ghg emissions increased by only 4%. From the 2000s to the 2010s, the number of natural disasters increased by 91%, while the average ghg emissions decreased by 4%. Over the four decades studied, the number of the various types of natural disasters is as follows;

| Decade | Storms | Winter Storms | Hurricanes | Droughts | Floods | Wildfires |
|---|---|---|---|---|---|---|
| 1980-1989 | 10 | 7 | 6 | 5 | 0 | 0 |
| 1990-1999 | 17 | 8 | 10 | 5 | 9 | 3 |
| 2000-2009 | 24 | 3 | 13 | 8 | 4 | 7 |
| 2010-2019 | 74 | 4 | 9 | 8 | 14 | 7 |

So there is no direct correlation between number or types of natural disasters in the U.S. and ghg emissions as listed in the previous paragraph. The purpose of this book is to investigate steps we in the U.S should take to save our natural planet. The major challenge is not in the U.S., but world wide. An estimate of global carbon emissions increase is given in a posting by *The Inimitable Tiff* accessed 2/29/20. (see Figure 1-2 below). Other references show similar increases, but the caution is that western countries have used East Asian countries such as China and India to manufacture goods such as electronics and clothing, and thus "outsourced" their emissions. Another factor to consider in this issue is the urbanization and agricultural clear cutting of forests in certain areas of the world, as explained in Book II.

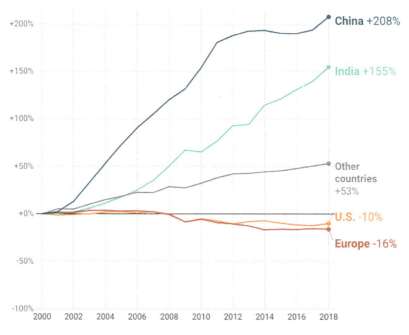

## Hurricanes Versus Carbon Emissions

CPI-adjusted hurricane events striking the U.S. mainland, and costs by decade (*NOAA Costliest U.S. Tropical Cyclone Table,* accounting for inflation to 2017 dollars, 1/26/18, updated through September 2019 ) have also been published. They are compared below with a cost per ton of GHGs emitted in the U.S (*Million Tons of Carbon Globally Emitted from Fossil Fuels,* NOAA Earth System Research Laboratory Global monitoring Division, 2018):

60s: 1 @ $1-$5 billion, 1@ $5-$10 billion, 1@ $10-$25 billion vs. 3000-4200 MTC/y (million tons carbon/year); 3 at an average of **$4.91/T**

70s: 2@ $5-$10 billion, 1@ $10-$25 billion vs. 4200-5200 MTC/y; 3 at an average of **$5.160/T**

80s: 1@ $1-$5 billion, 1@ $5-$10 billion, 1@ $10-$25 billion vs. 5200-6000 MTC/y; 3 at an average of **$4.383/T**

90s: 1@ $1-$5 billion, 3@ $5-$10 billion, 0@ $10-$25 billion, 1@ $25-$50 billion vs.

6000-7000 MTC/y; 5 at an average of **$11.80/T**

00s: 1@ $1-$5 billion, 3@ $5-$10 billion, 5@ $10-$25 billion, 2 @ $25-$50 billion, 0 @

$50-$125 billion, 1 over $125 billion vs. 7000-9700 MTC/y; 12 at an average of **$41.10/T**

10s: 2@ $1-$5 billion, 0@ $5-$10 billion, 3@ $10-$25 billion, 1@ $25-$50 billion, 4@ $50-$125 billion, 0 over $125 billion vs. 9700-10,500 MTC/y; 10 at an average of **$41.36/T**

These are the percent-average hurricane cost increases per decade per ton of carbon emissions: 5.1 percent, -0.15 percent, 16.9 percent, 24.8 percent, 0.63 percent. Like the previous graph showing natural disaster costs, there is no consistency with carbon emissions.

Both of these tables show an inconsistent increase in costs of these natural disasters per ton of carbon emitted in the U.S., but do not consider the relative cost or density increase of coastal structures built over the years, which would lower the relative cost of the most recent disasters.

The following graph plots these costs in billions of dollars for each weather and climate disaster, with over one billion dollars of damage from 1980 to 2017, versus millions of tons of GHCs (greenhouse gases) in $CO^2$ equivalent tons emitted in the U.S., and versus ppm of carbon dioxide. The graphs shows that, according to NOAA data, unlike the global warming question, there has been no close relationship between either carbon emissions or $CO_2$ concentrations and the cost of natural disasters over this period of time.

So the emotional issue of human-caused climate change must so far be considered false.

## Weather and Climate Disaster Costs Versus Carbon and CO2 Emissions

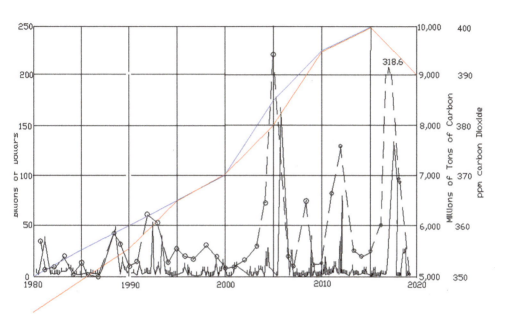

## Figure 1-3

The dashed lines and circles in this graph represent the accumulated total of natural disaster costs per year. The high peaks in 1992, 2004, 2005, 2008, 2011, 2016, 2017, and 2018 were caused by hurricanes, the largest of which was Katrina ($167.5 billion) in 2005, and Harvey ($130 billion) in 2017. The peaks in 1980, 1988, 2002, 2012, and 2013 were caused by droughts; in 1993, 2001, 2012, and 2016 were caused by flooding. In 2017 and 2018, they were caused by wildfires and, in 2011, were caused by tornados and a heatwave. The solid line represents each individual natural disaster's cost. The blue line represents U.S. Carbon Emissions in millions of equivalent tons, and the red line represents U.S. Carbon Dioxide emissions in ppm. For 2010-2017, disasters included 73 storms, 13 floods, 8 hurricans, 8 droughts, 7 wildfires and 4 winter storms. These numbers don't add up to those mentioned previously since they were from different sources.

## Does Human Activity Cause Melting Ice Caps and Rising Sea Levels?

There has been much discussion recently concerning rising sea levels and whether they are caused by melting ice caps due to global warming. Here is some data to help with that question:

- The total area of the earth is approximately 510,082,000 square kilometers. Of that area, the oceans cover about 70.9 percent or 360 million square kilometers.

- Using these figures, an analysis by Antarctic Glaciers.org reported a 0.0315-inch sea-level drop in 1992 and a 0.0709-inch sea level rise in 2016. At the 2016 rise rate, the sea level should increase 1.77 inches in the next 25 years, possibly because of global warming. If you assume that the maximum sea level increase in the next twenty-five years will be the 2016 rate plus the increase per year over the last twenty-four years, the total sea level will increase by 4.32 inches in twenty-five years.

- So, the sea level rise prediction should be somewhere between the current rate and the continued accelerated rate. Even this conservatively high water level increase does not seem to justify the fear expressed by many alarmists today, but still must be considered in planning and zoning decisions in the future.

## Solar Radiation Versus Global Temperatures

One argument against global warming has been that it is more closely related to solar activity than human causes. NASA, in its publication graphic: *Temperature vs. Solar Activity* shows a close relationship between global temperature rise following solar activity from 1880 until 1960, with a *rise* in temperature of 0.25°C, versus a solar irradiance *increase* of 0.8 watts per square meter. But a temperature *rise* from 1960 to 2019 of 0.50°C versus a solar irradiance *decrease* of 0.6 watts per square meter. NASA concludes: *It is therefore extremely unlikely that the sun has caused the observed global temperature warming trend over the past half-century.*

## Global Temperatures Versus Carbon Emissions

As explained above, the relationships between disaster costs and carbon emissions are not consistent. The relationship, though, between carbon emissions and temperature averages over the most recent fifty years is very consistent, as shown on the following graph:

## Average Global Temperature and World Carbon Emissions From Fossil Fuel Burning, (in millions of tons) 1880-2018

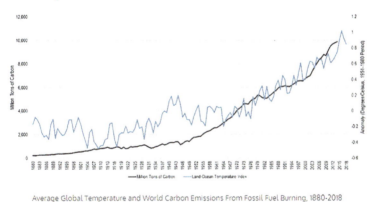

Average Global Temperature and World Carbon Emissions From Fossil Fuel Burning, 1880-2018

Source: Adapted from National Oceanic and Atmospheric Administration, Earth System Research Laboratory, Global Monitoring Division.

## Figure 1-4

On this graph, the carbon emissions are shown by the solid black line.

The primary weakness in this graph is that it does not reflect the heat sink of weather stations. In other words, since most weather stations are located in urban areas, over time, more heat-trapping structures are built in their vicinities. Therefore more heat is retained from warm days into cooler nights, increasing the average daily temperatures.

## Global Temperatures vs. $CO_2$

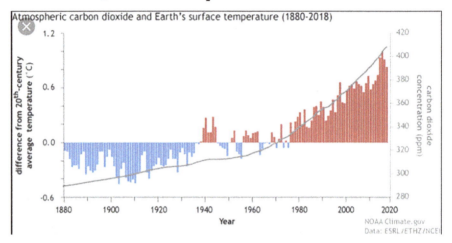

**Figure 1-5**

On this graph, the $CO_2$ concentration is shown by the solid grey line.

According to the *State of the Climate in 2018* report from NOAA, and the American Meteorological Society, global atmospheric carbon dioxide was 407.4 ppm plus or minus 0.1ppm in 2018, a new record high. That is an increase of 2.5 ppm plus or minus 0.1ppm from 2017, similar to the increase of 2.2 plus or minus 0.1ppm between 2016 and 2017.

In the 1960s, the global growth rate of atmospheric carbon dioxide was roughly 0.6 ppm plus or minus 0.1ppm per year. Over the past decade, however, the growth rate has been closer to 2.3 ppm per year. The annual rate of increase in atmospheric carbon dioxide over the past sixty years is about 100 times faster than historical natural increases, such as those that occurred at the end of the last ice age, 11,000-17,000 years ago.

It should be noted that the most important greenhouse gas in the atmosphere is water vapor. $CO_2$ has an effect of 1.3 to 4 percent of the water vapor effect. Still, water vapor has not been graphed in the upper atmosphere because of measurement inconsistencies (*Scientific Uncertainties, Why Scientists Disagree about Global Warming,* Idso, Carter and Singer, Nongovernmental International Panel on Climate Change, 2016).

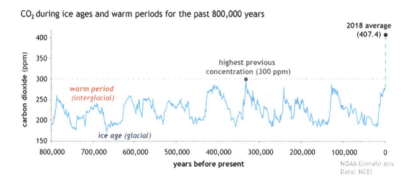

Atmospheric carbon dioxide concentrations in parts per million (ppm) for the past 800,000 years, based on EPICA (ice core) data. The peaks and valleys in carbon dioxide levels track the coming and going of ice ages (low carbon dioxide) and warmer interglacials (higher levels). Throughout these cycles, atmospheric carbon dioxide was never higher than 300 ppm; in 2018, it reached 407.4 ppm (black dot). NOAA Climate.gov, based on EPICA Dome C data (Luthi, D., et al., 2008) provided by NOAA NCEI Paleoclimatology Program.

# Figure 1-6

These last two graphs show a steady rise in surface $CO_2$ since 1975, and the last graph shows the estimate of $CO_2$ concentration over the past 800,000 years. In both graphs, there is an obvious increase that must be attributable to carbon emissions from human activities. But the deeper question is whether an increase in carbon emissions or carbon dioxide concentrations are related directly to global warming or climate change.

## CO2 Versus Altitude

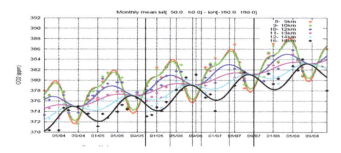

## Figure 1-7

This graph shows the $CO_2$ plotted versus dates from 2004-2008. It is apparent that all altitudes are close together in $CO_2$ in the fall and more spread out in the spring. Over a year, there is more spread than closeness. Therefore the atmospheric $CO_2$ increases less at the higher altitudes. Global warming is caused by ghgs in the ozone layer, so that altitude is the only one of interest in that concern. It has been found that $CO_2$ normally takes months to reach the ozone layer from the earth's surface, partially since $CO_2$ is heavier than oxygen.

The ozone layer is 9.3-21.7 miles (15-35Km) above the earth's surface and serves to absorb 97-99 percent of the sun's ultraviolet light, which otherwise would damage living organics on the earth. This layer contains less than 10ppm of ozone, while the average atmospheric ozone is about 0.3ppm.

## Temperature Distribution by Continent

An additional set of data available from Physics.org as well as other sources shows the temperature distribution for different continents. The U.S. and South America have a net rise in recent temperatures more along the East and West coasts. There is a drop in average temperatures in the center of the U.S. Canada and Russia have as much drop as the U.S. has rise. Europe has a rise all across the land, similar to the U.S. coasts. Asia has a lesser rise. Africa is mixed with a net rise.

Unlike the global warming graphs, these geographical maps show no consistent relationship between temperature and carbon emissions, apparently because of the global nature of climate and warming. In general, the more populated areas seem to have more temperature rise, as you would expect, because of the heat sink as described above.In addition, the oceans mostly have a drop in average temperature, and will affect coastal areas more than inland areas.

The Arctic has been warming over the late twentieth century as has the western Antarctic Peninsula, but the large polar Eastern Antarctic Ice Sheet has been cooling since the 1950s, according to a study; *Improved Methods for PCA-Based Reconstructions*, by O'Donald, Lewis, and McIntyre, *Journal of Climate 24*, 2010.

Ocean currents, temperatures, winds, etc., seem to have a greater impact on the geographical distribution of climate conditions and, therefore, require more study and emphasis.

## Summary of Record High and Low Temperatures in the U.S.A.

I have made an attempt to be more objective in this analysis than most sources, since you can basically prove anything you want with statistics. The following is a summary of the record high and low temperatures in the U.S. since 1960. The date most graphs show the beginning of the steep temperature rise, by region (National Climate Data Center, *State Climate Exteme Conditions*, 2015):

|  | # Record Highs since 1960 | # Record Lows since 1960 |
|---|---|---|
| E. Coast | 2 | 3 |
| W. Coast | 0 | 1 |
| SE | 2 | 6 |
| SW | 3 | 1 |
| NW | 4 | 7 |
| Alaska | 0 | 1 |
| Hawaii | 0 | 1 |
| Total | 11 | 20 |

The conclusion of the above analysis is that 35 percent of the records in the U.S. are hottest, and 65 percent are coldest. In addition, 82 percent of the record highs and 70 percent of the record lows are in the Southeast, Southwest and Northwest. In other words, the central part of the country, which is not affected by the oceans. As descibed above, this is the general trend for the rest of the planet. So, exteme cold temperatures have exceeded exteme warm temperatures in this country, since global warming supposedly began in 1960.

Is this result because of the heat sinks due to urbanization, since average temperatures from weather stations are going up? Is it because of greenhouse gases destoying the ozone layer? Is it inaccurate or uncorrected data, or is it an example of the misuse of statistics to prove a pre-conceived belief? The point of these questions is to emphasize the critical need for an objective scientific approach to this question before we continue to spend billions of dollars on a weak assumption. Those who ask these questions are accused of being anti-science, when the opposite is true.

## Conclusion of the Evidence of Climate Change or Global Warming

I have examined over one hundred graphs of historical global temperatures, and they are all similar in showing the "hockey stick" shaped rise in global temperature since 1960, but only a few attempts to show either NASA's or any other relationship between temperature rise and any possible cause.

These graphs virtually all use NASA and NOAA data as their source. However, the agencies have been very careful in not stating the reason for the rise in temperature, or whether that rise is related to human activity. Since the figures do not prove that increases in serious weather related disasters are caused by $CO_2$ or carbon emissions from human activity, but do indicate that global warming is affected by $CO_2$ and carbon emissions by human activity; data concerning *naturally*-caused climate change, must still be considered more carefully. We must concentrate our future efforts on global warming as it relates to human activity and not be sidetracked into an attitude of *climate change fear* without further evidence. It could be argued that global warming is climate change. But, if so, why not just admit it, call it global warming and get to work rather than assuming that whining will solve the problem?

Following NOAA's example of only reporting data, the summary of this available scientifically based global data is as follows:

- Global natural disasters over $1.0 billion: Cost vs. carbon and $CO_2$ emissions, inconsistent with time, have no apparent direct relationship between human activity and climate change.

- U.S. hurricane events: Cost vs. carbon emissions: inconsistent with time, there is no apparent relationship between human activity and climate change.

- Ice cap melting is increasing, due to global warming, and is estimated to cause global water levels to rise between 1.77 and 4.32 inches over the next twenty-five years.

- Solar Activity: Solar irradiance vs. global warming: inconsistent with time, therefore does not prove that solar radiation is the total cause of global warming.

- Temperature vs. carbon emissions: consistent with time without considering weather station heat sink effects or record low temperatures; favors, but does not prove global warming caused by human activities

- Temperature vs. $CO_2$: consistent with time; favors, but does not prove global warming caused by human activities, except for the effect of urbanization and the evidence of record low temperatures; but like carbon emissions, favors a relationship with human activities.

- $CO_2$ vs. Altitude: The $CO_2$ in the atmosphere and lower stratosphere has an increase with time in the upper and the lower altitudes. It is 6-10ppm higher at the lower altitudes than at the higher during Winter and Spring, and at equal or lower levels during Summer and Fall, but it takes several months for $CO_2$ to be carried from the earth's surface to the stratosphere. This increase in $CO_2$ with time in the ozone layer, along with a ban on CFCs apparently has counteracted the 1970s concern that chlorofluorocarbons (CFCs) were depleting the ozone layer. Since then, the ozone layer depletion has slowed or stopped, according to *Scientific Assessment of Ozone Depletion, Executive Summary,* 1914, NOAA. Any future studies in temperature versus $CO_2$ should be concentrated on the $CO_2$ in the ozone layer, like the NOAA study, and not on the surface of the earth, since the theory of global warming is the depletion of the ozone layer due to ghgs.

- Continental temperatures related to geography: inconsistent between coastal and interior, or north and south, favors warming relationship with urbanization

- High and Low Record temperatures in the USA: inconsistent with time, favors global cooling

**So, there appears to be no consistency between climate change and climate disasters or hurricanes, but there does seem to be a relationship between global warming and altitude, ocean effects, urbanization, carbon emissions and $CO_2$, but not solar activity.**

Remember that there is a consistent mass balance of water, in the world, so the damage from water occurs due to the uneven distribution of water, not the total amount. Hurricanes form near the equator where the oceanwater is warmer, at least 80F. The water evaporates due to winds, and rises, forming a moving cloud. The east to west trade winds, carry these storms from their genesis, usually near Africa, to areas such as the Caribbean, Mexico and the Southeast U.S. The cycle of evaporation and cooling as the water vapor rotates in the cloud forms a tropical depression at wind speeds from 25-38mph, a tropical storm at 39-73mph and a hurricane above 74mph.

Again, this process transfers water, and does not create water. It would be virtually impossible to alter trade winds, but there is a possibility of lowering tropical ocean water temperatures by reversing global warming. As a result of the lack of scientific facts reported by these studies concerning climate change, we direly need to go several steps further to determine the, if any, relationship between global warming, climate change, and human activity. As much as the media and vested interest groups would like to tie the three together, it simply hasn't happened.

I believe that our interest instead, should be exclusively on the logical and scientific facts of our continued pollution of the environment and our use of limited natural resources, with emphasis on carbon and $CO_2$ emissions. Then we should be able to reach conclusions and develop responses to really save our planet, rather than politicizing the issue and accomplishing nothing.

In consideration of how future research should be accomplished, we must consider the "stakeholders" in the issues of climate change and global warming. It seems that the very questions posed in this book involve these same stakeholders:

- The inhabitants of this planet

- Those preceding us who have made advances in the fields of science, philosophy, and theology

- Agricultural, municipal, commercial, industrial, and personal polluters

- Engineers and scientists responsible for minimizing or eliminating this pollution

- Government officials responsible for regulating pollution

If we analyze the boundaries of our responsibility and the tools we have to alter present practices that may be causing damage, we should consider the following:

- Our boundary of study is our planet and, potentially, the rest of the universe.

- There are at least two possibilities of the demise of our planet: it is polluted to the point where human habitation becomes impossible and/or we will use up all of the non-renewable natural resources required to sustain civilization. We will explore the first possibility in Book III, the second in Book IV.

- There are other possible methods to end our planet, such as through war, which are beyond the scope of this book.

- There is another possibility that we should consider if we are to truly open ourselves up to all views of the future that are practical. That is that God has relegated our existence to be temporary due to our non-acceptance of His earlier creation, which was intended to be permanent (Genesis 1:31).

- The only real tool we have to affect the life of our planet is our cumulative minds. Book II will describe the relationship between our minds, our brains, our souls, and our spirits, and attempt to determine the capability of these tools to save our planet.

- Since history has momentum and is the source of all knowledge (see Chapter 2), our tools include the accumulated knowledge of history on how our minds work to determine the limits of our capability to make and carry out the decisions which may be required to save our planet.

## Is There Both a Natural and a Spiritual World?

The next question we will address is whether the four mental tools we possess – of *brain, mind, soul, and spirit* – are real. If so, can they be harnessed to allow the implementation of policies and procedures which will accomplish our goal of sustainable growth? Concerning these listed tools, at this point, if we wear blinders so that we accept only the natural set of tools or only the supernatural set of tools, we will never understand, or even have the opportunity to consider, what constitutes the complete human existence, capability, or capacity.

God has placed us in an interesting position, as part of a permanent spiritual world, and at the same time, part of a temporary natural world.

The spiritual world is only more complex because we can never completely understand it. God gave us five senses to understand the natural world, but only one to understand the spiritual world. And that one is the essence of spirituality, the Holy Spirit; God Himself in those of us who accept Him. The access to the Holy

Spirit and the understanding of the spiritual reality is only through belief and acceptance of Jesus Christ as our Savior from sin. Only those humans who have a belief and acceptance in the Creator of humanity can have the right to penetrate that barrier between the natural and the supernatural. Simple and free to everyone, but impossible for those who see themselves as having limitless power as their own god. The only communication with the spiritual world is God's written Word for us in the form of the Bible. That is literally the only written document that claims, within its words, that it is entirely spiritual in nature and provided directly from God. So, I'll let that speak for itself; I can't compete with it.

The natural world, even though much less complex, is the one we face every day; and seemingly is terribly complex on its own. It's really not, and since God has placed us there, it is our job to find our way through it in a way that pleases Him and sustains us.

Much of our confusion is caused by some of our theoretical scientists, such as Charles Darwin, Carl Sagan, and Richard Dawkins, who are simply science fiction writers, and should not be considered legitimate scientists with true facts to communicate. Rather, they are only entertainers. These scientists are not qualified to speak of God's Creation, His existence, or His control of the natural world since they have no ability to communicate with or understand the spiritual world. They can speak of the natural world, but their work becomes questionable when they move beyond their expertise in science. If they are true scientists, they should remain in their discipline, and not be confused into believing and claiming that theoretical science and the supernatural meet the tests of science. They should admit that their beliefs are an imaginative theory, which brings no value to true science, especially when it involves theology.

Our part in this issue is to be able to differentiate between natural fact and fiction, and to open our minds to a reality outside of that which we can measure and understand scientifically.

I challenge you to use your intelligence through logic, reason, and critical thinking to really understand the organized universe, to differentiate between the imagination of evolution and life in outer space, and be able to comprehend the logic of integrated reality. The sciences of mathematics, archeology, and chemistry have largely disciplined themselves to nontheoretical conclusions, while physics, biology, astronomy, and geology have allowed the consideration of unproven theory and imagination to be considered as legitimate science. Many major universities are offering courses and even majors in "evolutionary science," a genuine oxymoron.

Why do we not have the honesty to force scientists to follow the laws of science and confine their conclusions to observed, documented, and repeatable facts? Why does macroevolution get a "pass" from true science in order to salve the reputation of its science-fiction believers? Why not challenge them to show us a fossil of a "bat/rat," an "ape/human," a "bird/man," etc., which is not one or the other with its appropriate DNA?

To be objective, we must ask the same questions of religion, but the simple answer is that science is not applicable in theology because science is limited to the natural, while theology is involved with the supernatural. We should never consider imagination in natural science, while only imagination and faith *can* explain the spiritual. Reality consists of both realms, which this book will attempt to explain.

## The Natural World as Experienced by Our Senses

Since our perception of the natural world is through our five senses (see below), and since they are responsible for 100 percent of the memories stored in our brain, I'll start there.

Our understanding of the natural world, which is our environment, through our senses, at its simplest level involves, initially, the inorganic portion of the natural world. More complex is the organic portion of the natural world. Much more complex is the living organic portion of the natural world. And extremely complex is the human portion. Let's go there now, for the lesser portions are relatively easy to find our way through.

In every interaction with another human, we leave a memory through their senses. Is the permanent memory of us positive or negative? "Positive" being right, true, encouraging, instructive, and just. "Negative" being wrong, untrue, discouraging, instructive, nevertheless, and unjust. Who defines these terms? That is pretty obvious. God, or if we don't believe in God, ourselves as our god.

So, in every interaction with another human, we leave permanent memories of ourselves which may change them or not. In effect we are limited to our sensory experiences since they are the total content of our brains. We can manipulate those memories and arrive at unique conclusions, but the facts remain limited to experiences. Beyond is imagination, which can be the basis for science but is not science, only a theory until it meets the definition of science.

Let's look at sensory interactions now in order to demonstrate what memories fill our brains:

- **Taste**: Let's not go there.

- **Touch**: Ideally limited to intimacy or friendship.

- **Smell**: Only good or bad smells seem to last in memory as a general point, and they are typically a matter of chemicals, hygiene, or natural deterioration. For the following senses, I will go into some detail and personal reflections in order to demonstrate what our memories, and therefore, our brains and minds consist of.

- **Sight**: The way we look is our first opportunity to communicate with others. Do we try to look just like everyone else? If so, the "look" is ever-changing. In order to keep up and conform, we must diligently study the trends of society and discern what is lasting and what is a temporary fad.

Or do we want to look different from everyone else, or be known as a trendsetter, or leader? If so, why? Think that through. What do we want to look like in the eyes of others? Smart, sophisticated, tough, strong, athletic, cool? Do we want to look like our vision of others in our past? A "braino," an aristocrat, a Mafioso, a hippie, a jock, a cowboy, a forward thinker, or just what is the character we are determined to be? Are we just dressing ourselves up with no place to go? Are we just decorating ourselves temporarily, or permanently, in a way that will seem foolish to others and even ourselves at a later point? How far should we go in playing this role, and what is its effect on our audience? So, our appearance will affect the sensory perceptions of our contacts, positively or negatively.

Our remaining sight sensory input is typically related, not to the perception of our communicator, but what they do, whether they be fellow humans, animals, or organic or inorganic sights.

- **Hearing**: What sounds do we implant permanently in others in the form of memories?

Do we think that out? Are they truly what we want to be remembered for? Unlike sight, significant memory inducing sounds are usually from humans.

I am cursed with a kind of photographic memory, and in one way or another, everybody is. So what do we want others to remember about us?

Or are your "sounds" pretty much just limited o social media in order to give you one less sensory output to worry about? In the Bible, Jesus' blood brother, James, says that the tongue: "*...is a restless evil and full of deadly poison*" (James 3:8) and "*From the *same mouth come both blessing and cursing. My brethren, these things ought not to be so"* (James 3:10).

So, there seems to be no question that our sensory input to others, with our actions, our tongues, or in writing, is the most impactful to others, positively and negatively, in the natural world. That should be a caution for us as we grow and mature. Everyone we contact from our friends to our family, to our wife and children, is changed a little, or a lot, by the words we say to them, or to others in their presence, just as we have been changed through our lives by the sights and words of others.

Therefore, if we realize that these sensory perceptions may be permanently lodged in our memories, even subconsciously, we should monitor to whom we speak, listen, read, and observe.

## The Spiritual World

Now, what about our position in the spiritual world? We can disbelieve that it exists. To some, that is an easy way out. If it doesn't exist, they don't have to deal with it or understand it.

**The problem, logically, is that no thinking person should say something does not exist, especially if it is supernatural, since it can't be proven or disproven, anyway**.

Living a life of disbelief is futile and illogical since disbelief means nothing, especially when it comes from the inexperienced.

Most people in the U.S. do accept the possibility of, or believe in some form of supernatural existence. Pew Research Center, October 9, 2018 says 83 percent of the U.S. population believe in God, another 12 percent believe in the supernatural and 7 percent are atheists.

Those who believe in some form of the supernatural must imagine their beliefs with their natural minds based on sensory experiences and logic, or must have a spiritual connection to the supernatural itself.

That connection has only been claimed and documented through the Bible, or by God through certain chosen humans. The logical weakness of religions claiming the latter is that their documents claim to have been produced by humans, rather than God, as the Bible claims.

The only way to discern spiritual truths then, is from the Bible, as the only book claiming to be inspired entirely and completely by God. The "ologys" of humans reveal nothing of the knowledge of spiritual things because they have no source of experience. The famous leaders of science, philosophy, and even theology have little to offer us spiritually. They are not part of the Gospel Story, which is exclusively about the triune God. It only incorporates man as a creature of creation, until that creature has accepted the gift of the spiritual knowledge available to him through God as the Holy Spirit. Another way to look at it is that man is no more part of the story than a radish, until he becomes part of the church, the kingdom of heaven (see Chapter 16). So, there is no reason or value to worshipping, or learning from leaders, authors or preachers, unless they are part of the *Church Universal* and are acting as such by teaching from the Scriptures.

## How Does the Bible Fit In?

In order for God's children to discern spiritual truths, He has, in His perfect love, provided us with a written communication in the form of the Holy Bible, which claims infallibility.

Is it reasonable to assume that a perfect and all-powerful God would see that His only communication with His children would be inerrant in its original form? Is it beyond the ability of a perfect and all-powerful God to assure that His communication product is preserved and translated throughout millennia without confusion and contradiction? I conclude that the answer to the first question must be *yes*, and the answer to the second must be *no*. On the contrary, it would be inconsistent for a perfect God to not do everything, including communicating perfectly.

## Using Our Humanity to Think

Throughout history, philosophers, theologians, and scientists have debated the constituents of human existence. The *Trichotomy* view of Plato was that man is made up of the physical *body*, the *soul*, and the *spirit*.

Today, with our understanding of the brain, and to a certain extent, the mind, a body/brain/*mind*/*soul*/*spirit* differentiation is more relevant. The conundrum of understanding both natural and supernatural reality is that we must be wise enough to fathom the natural, and we must be as simple as a child to fathom the supernatural. There is no *natural* education or experience that will aid in our spiritual growth, but since the supernatural created the natural, as proven by the Law of Conservation of Matter, the natural reality can indeed be better understood through a spiritual connection.

Virtually all of the scientific laws we follow today were discovered by Christians or Jews who submitted to a higher power they called *God*. Their view was that nature revealed God.

As an environmental scientist, I have found that the environment can only be completely understood interdisciplinarily. Likewise, as an elder and Christian teacher, I have found that spiritual reality can only be completely understood in the same way. Indeed, a child is not constrained by disciplines.

The lethal mistake that many academics make is by restricting their research and study to their own limited disciplines; scientists to their scientific specialty, theologians to the work of other theologians, and philosophers tend to make the same, limiting mistake. Ironically, the second lethal mistake is speaking personally outside of academic and educational experience, rather than quoting experts. There is only one Being who can completely understand the supernatural part of reality, and that is God, who uniquely has been present in the supernatural forever. That God who can indwell His human creation in the form of the Holy Spirit is our only hope of growth in the spiritual realm.

With the exception of the only document created by God, the Holy Bible, all other worldly knowledge, logic, writings and other communication are *relatively* spiritually worthless, since it is only a natural opinion. In other words, the Holy Spirit in us interprets the Scriptures for us, and without that interpretation, we cannot understand the Scriptures. Humbling isn't it?

So, this book seeks to examine the human mind from a natural *and* supernatural, or spiritual perspective, with the purpose of discerning whether that mind is capable of saving the planet. God has only created one natural being potentially capable of spiritual thought and discernment, and that is a human.

**If there were no humans on the planet, there would be no possible spiritual awareness**.

The only logical reason for that ability is that God has appointed a limited number of humans to be His children, in which to bodily house Himself. Yes, God can choose to do what he wishes since He is all-powerful. He, in His love, has granted humans free will, and since we were created in His image, He also possesses free will, but is not subject to sin as we are. He, rather, is perfect, fair, and just.

According to the Bible, the only things God cannot do is lie, be tempted by evil, and deny Himself.

## Thinking About the Environment

For the above reasons, I feel called to present the results of my research into the human mind, and its capability to save our planet. It seems that much of the current controversy over environmental control is an emotional approach to the belief or lack of belief in a Higher Power. This limited thinking is not the proper process to use in making the monumental and very expensive decisions that must be faced if we are to survive as a country or a world.

The natural environment cannot be understood by only one discipline of science. I suggest questioning the authority of scientists who pontificate on disciplines in which they are not qualified, especially in theology. Their decisions in these unqualified arenas are relegated to emotion and not true science.

I likewise suggest questioning the authority of theologians who pontificate in disciplines in which they are not qualified, especially in science. They can and should speak theologically and even philosophically, but without appropriate education, they are not qualified for scientific judgments.

Therefore, in this book, I try to reference all information outside of my personal experience and education and invite the reader to research these and other qualified scientific and theological sources. To encourage further research, I have placed references immediately following their context.

There is no attempt in this book to change the pronouns as presented in quotes from the Bible or comments thereon. The reader should realize that all of God's promises and references are to women in the same way as to men. Woman was made from man. Therefore, they are spiritually identical and made in God's image. In addition, many of the pronouns in the Hebrew and Greek manuscripts are written in the neuter, but translated in the masculine.

# BOOK II: WHAT IS THE SCIENTIFIC MIND?

*Chapter 2*

# Existence And Reality

## Introduction

In order to rationally answer the challenge posed in the title of this book, we should define and differentiate *existence* and *reality*. We should also determine if there is an intrinsic difference between the four basic types of natural existence: inorganic, non-living organics, living organisms, and humans. Only then can we begin to understand and to answer the challenge: *How to Save our Planet.*

Some of the information contained in the sections on existence and reality is taken from [8]*The Future of Species,* by this author, E. Roberts Alley.

## What is Existence?

Virtually everyone, at some time during their life, faces the critical questions of our existence. Why am I here, at this time in history, at this point in space, as a part of this family? Why am I solid matter and not just a thought or imagination? Why can science explain our bodies, including our brains, in terms of molecular structures, but not be able to understand our minds? Why can't science tell us how inorganics, organics, and living organics came into existence originally, when we know how plants and animals are able to grow over their lifetimes and create mass.? If we do not carry our curiosity back to the logical source of our existence as humans, we will have difficulty accepting our place in civilization. We will become self-centered and isolated. Is this the problem with society today? Have we become navel-gazers with no realization or appreciation outside of our own interests?

**In order to prevent this natural tendency of selfishness, we must face these eternal questions, or we will become attracted to nihilistic world views such as secularism, atheism, Marxism, postmodernism or new spirituality.**

We understand that living organic growth stems from the splitting of living cells, gradually adding weight to a plant or animal as it ages. But how did these cells learn to reproduce by division, and turn an acorn into a thousand-pound tree, and a mother's egg into a hundred-pound human? Only an intelligent source, even more, only a Creator, could negate natural evolutionary theory, create DNA, program growth and inherited characteristics, but not allow its highest product to exist forever, as do the lower products.

Why would an all-powerful Creator allow His ultimate creation to be limited by death? Initially, according to the Bible, He did not. He created a "good" world and universe and a "very good" human being, which had DNA, but no capacity for death (Gen. 1:31, 2:17, 3:19, Rom. 5:12, 15-17).

But that human, as have all humans since, had the weakness of wanting to understand his Creator – indeed, to replace his Creator with himself. If we truly have to understand everything that God has done, our pride leads us to the point that when we, or some other human, cannot understand something, we are forced by our arrogance to accept that it does not exist. At that point, we want to be God. That selfish desire is impossible to realize, since by definition, God is the only entity that has existed forever, and is the only explanation of the source of the universe. The only just penalty that God could have for His Creation wanting to take His place is death, in order to limit the sovereignty of humans. Then we could not be omniscient, omnipotent, or omnipresent like God.

At that point, there was no death in the created world. The only entity which could have allowed death into the Creation is the Creator of life Himself. Along with the fall of man, God also justly cursed all other living organic matter, vegetation, and animals, since they provided man's sustenance.

For natural life, the opposite of death is health – or life. Health is the balance between internal physical life and external nature. Everything outside of our bodies works toward our death. Organisms which allow us to live, destroy us after death.

The only solution then, to death, the problem of existence, was for God to personally enter our universe as a living organic being in the form of Jesus Christ. Jesus was fully human and fully God, incarnated – as was the whole universe – from nothing but God, and as God, to communicate to us, His Creation, His plan to return to us the opportunity of the original creation, which is to exist forever.

This huge question of our acceptance of the supernatural must be answered if we are to understand the possibilities for the future of our planet, including how the natural and the supernatural relate. The overwhelming majority of the population has some belief in the supernatural. It may be belief in New Age, re-incarnation, mental telepathy, ghosts, Hinduism, Buddhism, Islam, Judaism, or Christian, but it is based on spirituality as reality. The ability to recognize that reality is not limited to natural occurrences but is the ultimate key to understanding our relationship to our past, present, and future.

Many of us, as did Darwin, choose to base our beliefs primarily, or solely, on the findings of science. Natural science is limited to a study of the natural, which is comprised of those things in our universe that we can sense through sight, hearing, smell, taste, or physical feeling. Natural science is incapable of analyzing the supernatural or spiritual.

**If there is only science, and no supernatural, there is no God. If science is our god, the accumulation of human knowledge is the limit of our truth, and our Bible, and changes every day.**

When we exclusively, through our efforts and intellect, attempt to discount, or to explain the supernatural, we will become frustrated because the supernatural cannot be completely understood by the natural mind. Spiritual truth is not learned by intellectual reasoning. It is learned only through the Word of God in Scripture as interpreted by the Holy Spirit. Hard, foreign-sounding words? Yes, but that is the nature of God. He is different from us, unnatural to the human mind, but completely true, fair, and just, and therefore, the only realistic standard on which to base our lives.

And just because we love, respect, and understand God's created natural universe, it doesn't mean that we are left out of the ultimate spiritual truth. The first visitors from afar to worship Jesus were the magi, a cast of wise-men, specializing in astrology, medicine, and natural science (Matt. 2:1,2).

## What is Reality?

Throughout history from Socrates and Plato, through St. Augustine and Thomas Aquinas, to the present, philosophers, theologians, and later, scientists, have attempted to categorize human reality. One question over the ages has been whether reality includes a supernatural element, or whether it is limited to the natural.

My personal belief is that reality is present in six forms: the omnipotent God the Father, the *I AM*; God the Son, the Messiah, the sacrifice for our sins, the Word; God the Holy Spirit, present in us as Jesus' legacy; ourselves; others; and nature.

Reality is either order or disorder; the order of God's Creation, as the gift of science, or the disorder of evolution, as the random accident of change. And life is either destined for the ultimate order of Heaven, or the ultimate disorder of Hell. The most consistent and accepted categorization in literature of reality seems to be as follows:

## Natural Reality

*Body*: The physical substance of a human containing DNA.

*Brain*: That part of the central nervous system of the body, within the cranium, in other words, the organ of thought, memory, and emotion. It contains all the higher centers for various sensory impulses, and it initiates, controls, and coordinates muscular movements. The function of the brain has only been understood since the early 1900s AD.

*Mind*: The seat of consciousness, feeling, and will, the intellect. (*New Webster's Dictionary and Thesaurus of the English Language*, Lexicon Publications, Inc., Danbury, CT, 1992), Old Testament: "heart, breath"; New Testament: "deep thought, the intellect." Twenty different Greek words are translated as "mind."

## Supernatural Reality

*Soul*: The immortal part of man, as distinguished from his body. As defined by *Webster's* (ibid). Old Testament: consistently: "breath." New Testament: consistently uses "breath, spirit.

*Spirit*: The intelligent or immaterial part of man as distinguished from the body (Webster's, ibid). Old Testament: consistently: "wind, breath, spirit." Not capitalized in the Hebrew. New Testament: consistently: "Current of air, breath, spirit." Not capitalized in the Greek.

## A History of our Effort to Define Existence and Reality

The principle disciplines which seek to understand the way that we think, incorporate, besides our bodies, the four mental tools possessed by humans; *brain, mind, soul, and spirit*. The categories of existence and reality have been debated throughout history, especially as they relate to these mental tools. Sometimes we need to understand our history in order to understand our present and our potential future.

The following sections are not intended to be comprehensive; they are rather intended to show the relative progress over the last several millennia as science, philosophy and theology have matured, in order to demonstrate how the thinking and reasoning of the intellectual world has progressed; and how it has not. Hopefully, that will encourage us to make a further leap into the future in order to realize sustainable growth on this planet.

## A Brief History of Neuroscience

The following information is presented as a general summary of the history of neuroscience:

- 2050-1710 BC: In ancient Egypt, from the late Middle Kingdom onwards, the *brain* was considered "cranial stuffing," and removed prior to mummification. The *heart* was assumed to be the seat of intelligence. Wikipedia, *History of Neuroscience*.

- Historians typically will not admit it but, when listing the history of neuroscience, in the book of Genesis, which was written between 1500 to 1300 BC, the word *heart* is used for the word *mind* many times (Gen. 6:5, 6; 8:21; 17:17; 20:5, 6, etc.).

- 600 to 500 BC, the Pythagorean Alemaeon of Croton first considered the *brain* to be the place where the *mind* was located. Wikipedia, ibid.

- Aristotle, 382BC-320BC, thought that while the *heart* was the seat of intelligence, the *brain* was a cooling mechanism for the blood.

- 400-1000 AD, Dark Ages. During the Roman Empire, the anatomist Galen noted that specific spinal nerves controlled specific muscles (ibid).

- 475 AD, Fall of the Roman Empire

- 475-1549, Middle Ages

- 936-1013 AD. Early basic surgical neurology by Al-Zahrawl

- 1449-1650, The Enlightenment, Renaissance, Age of Discovery

- 1596-1650 AD, Descartes, attempted to relate the *brain* to the *mind* through Dualism

- 1632-1723 AD, Anton van Leeuwenhoek, the Dutch merchant who discovered the microbial world. He disproved the before held belief in spontaneous generation. Wikipedia,ibid.

- 1822-1895, Louis Pasteur confirmed that micro-organisms do not arise spontaneously.

- 1875-1906, It was not until the second half of the 19th Century that experimentation on animal *brains* allowed the discovery of the role of electricity in nerves and the existence of neurons. Note the gap in neurological advancement from 475 to 1875.

- 1920-1935, discoveries which demonstrated the similarities of all living systems at the metabolic level, thus unified the sciences of biochemistry and microbiology.

- 1944, the discovery by Avery, McLeod, and McCarty that the process of bacterial genetic transfer is mediated by DNA, Wikipedia,ibid.

- 1953, publication in *Nature* of a paper by J.D. Watson and F.H.C. Crick proposing the structure of DNA, Wikipedia,ibid.

- Mid-70s, discoveries in *Recombinant DNA Technology* which merged chemical and biological studies on genetic material so that DNA can be manipulated *in vitro*

- In the 20$^{th}$ Century, the *brain* was categorized into specific areas responsible for specific tasks.

- 2019: Rapid scientific advancement in microbiology using DNA in the fields of pharmacology, genetic engineering, disease prevention, agriculture, forensics and archeology.

- Recent 2019 blogs continue to debate the two polar opposite beliefs of dualism and materialism among neuroscientists. A dualist differentiates between the *brain* and the *mind,* recognizing that the *brain* function cannot explain consciousness. The *brain* is an organ; the mind isn't. The *brain* is a physical place where the mind resides. It is a vessel in which the electronic impulses that create thought are contained. With the *brain*, you coordinate your moves, your organism, your activities, and transmit impulses. But you use the *mind* to think. The *mind* is the manifestations of thought, perception, emotion, determination, memory, and imagination that takes place within the *brain*. *Mind* is often used to refer to the thought processes of reason. The *mind* is the awareness of consciousness, the ability to control what we do and know what we are doing and why. It is the ability to understand. Animals are able to interpret their environment, but not understand it, whereas humans are able to understand what happens around them and adapt.

## A Brief History of Philosophy

As the quest for an understanding of how we think has progressed, along with the early scientific advancements based on a Creator God, and the later thoughts of materialism, scientists began to concentrate completely on the natural or material aspects of our thinking, and philosophers concentrated on the immaterial aspects. But the early basis for science was consistently the existence of God.

- 1800-300 BC, Cultural beliefs based on legends which supposedly explained natural phenomena.

- 500-300BC, Natural Philosophers, transition from unconfirmed mythology to reason and critical thinking.

- 430 BC, Socrates, an Athenian, born 469 BC and martyred for his teachings in 399 BC, began lecturing in Athens, Greece. He wrote nothing, and we know about him only from the writings of his pupils, Xenophon and Plato. Socrates made a profession of no knowledge except his own ignorance, the famous "Socratic Irony."

- 380 BC, Philosophically, these ideas were called Metaphysical Dualism and were popularized by Phaedrus Plato, 427-346 BC. Plato was a pupil of Socrates from age twenty. Socrates, according to Plato (*The Apology of Socrates*, *The Harvard Classics*, P.F. Collier & Son Corporation, New York, 1909, 1937) said, "*Men of Athens, I honor and love you; but I shall obey God rather than you...*"

- Plato was born in Athens Greece and established a school of philosophy in Athens called the Academy about 388 BC. He taught that the physical and intelligible worlds exist simultaneously and that the physical world is but a shadow of the intelligible immaterial. He said that good and evil are states of the immortal *soul* and that the *soul* is divided into:

(1) The *rational* part, our ability to judge, love and discern truth, and ideally rule over the other parts by *reason;*

(2) The *spirited* part, our emotional ability to feel empathy toward honor and victory; and

(3) The *appetitive* part, our desire for food, drink, and sex (Wikipedia, *Plato*).

Good and evil were part of the immortal *soul*. The *physical* world to him included senses, images, and objects, prone to change, and based on opinion. While the *intelligible* world included intellect, thought, and forms, based on knowledge, and not subject to change.

- 350 BC, Aristotle, 382-320 BC, a pupil and philosophical rival of Plato, wrote that the *soul*, or psyche, is the form, or essence of any living thing, the perfect expression, truth, meaning, or realization of a natural *body*. He identified *reason* as the *soul*, with God as the eternal, omnipotent *thinker*.

Aristotle considered the *soul* as indistinct from the *body* that it is in, and that it is the possession of a *soul* of a specific kind (vegetable or animal), that makes an organism what it is, and thus that the notion of a *body* without a *soul*, or of a *soul* in the wrong kind of *body*, is simply unintelligible. He regarded the *soul* of a plant as *nutrition*, the *soul* of an animal as *movement*, and the *soul* of a human as *reason*. He believed that these characteristics are cumulative; that is, the animal soul is characterized by *nutrition* and *movement*, and the human soul is characterized by all three.

Aristotle seemed to think of the *heart* as we would the *mind*; as a unity, as the common sense organ, to organize all senses. He argues that some parts of the *soul*, such as the intellect, can exist without the *body*, but most cannot. (Wikipedia, *On the Soul*).

- 323 BC-420 AD, Hellenism, The decline of Greek culture and the fusion of multi-cultures

- 400 AD-1000 AD, Dark Ages, Emphasis on Christian theology

- 475-1549 AD, Middle Ages

- 1449-1650 AD, Enlightenment, A time of rapid cultural, religious and politicaladvancement

- 1596-1650AD, Rene Descartes, the originator of the phrase, "I think, therefore, I am."

- 1543-1727AD, The Scientific Revolution, an emphasis on rationalism, leading to empiricism, a reliance on observation without regard to theory or science. Promoted rationalism rather than empiricism. Believed in the absolute freedom of God's act of Creation.

- 1694-1778, Voltaire, Promoted freedom of religion, separation of the church and state, and was critical of the Roman Catholic Church.

- 1724-1804, Immanuel Kant, promoted transcendental idealism; things exist, but their nature is unknowable. He argued for the reason and truth of free will, God, and the immortality of the soul, but against many conventions of the Church.

- 1818-1882, Karl Marx, political theorist and socialist revolutionary.

- 1844-1900, Friedrich Nietzsche, spoke of the "Death of God," foresaw the dissolution of traditional religion and metaphysics, embraced nihilism, self-realization, power as being good, weakness as being bad. His writings were contradictory and very inductive in reasoning.

- 1856-1939, Sigmund Freud, founding father of psychoanalysis.

## A Brief History of Theology

Some of the information in this "Brief History of Theology" is taken from *The Story of the Bible*, Larry Stone, Thomas Nelson, Nashville, TN, 2010.

- 4004 BC, Creation, according to Bishop Ussher, (1580-1656 AD), took place on October 22, 4004 BC. Chinese literature estimates as long ago as 3400 BC. Egyptian literature estimates 3110 BC; Babylonian, 3300 BC; Jewish, 3762 BC. Close for the major religions of the world in their separate estimates. God established a unilateral covenant with His Creation of life and dominion over the earth (Gen. 1:27-31, Jer. 31:35-36).

- 1813-1638 BC, Abraham, the common patriarch of the Jewish, and later the Christians, and then the Islamic religions. God established a unilateral covenant with Abraham as a promise of land, seed, and blessing (Gen. 12:1-3).

- 1392-1272 BC, Moses, the prophet who led the Israelites out of slavery in Egypt and into the Promised Land of Canaan. Moses received the Ten Commandments from God and wrote the first five books of the Bible, known as the Pentateuch. Moses, many times, called God *Yahweh*, in Hebrew, translated as *I am the Lord*. God established a unilateral covenant with Moses of laws and rules (Jer. 31:33, Ez. 36:27, 37:24).

- 1039-969 BC, King David, the king of the United Kingdom of Israel and Judah, after Saul. David was a "man of God" and wrote many of the Psalms. God established a unilateral covenant with David as king of Israel and prince forever (through Jesus, Ez. 37:24-25).

- 931-425 BC, the major and minor prophets wrote the last seventeen books of the Old Testament

- 563-483 BC, Buddha, the founder of Buddhism.

- 350-140 BC, Translation of the Septuagint from Hebrew into Greek

- 0-30 AD, Jesus Christ. After the life of Jesus, most thinking revolved around His radical claims. Even though the excesses of the Roman Empire destroyed many historical documents during this relatively stagnant scientific and philosophical period, theology, based on the teachings of Jesus Christ, and later aided by the establishment and intellectual progress in the Roman Catholic Church, Jesus captured the reasoned imagination of the Western world.

- Mid 40's AD-late 90's AD, books of the New Testament written

- 60 AD, Dead Sea Scrolls hidden

- 125-200 AD, Old Testament Papyri discovered

- 27 AD-476, Roman Empire

- 200-250 AD, Septuagint Old Testament written by Origen, a translation from Hebrew to Greek

- 320 AD, Codex Vaticanus written, the oldest nearly complete copy of the Septuagint Old Testament Greek Bible and the Greek New Testament. The source of the Alexandrian family of manuscripts.

- 400 AD, Latin Vulgate Bible, the translation of the Old Testament into Latin

- 570-632 AD, Mohammed

- 400-1000 AD, Dark Ages

- 350-430 AD, Augustine of Hippo

- 475 AD, fall of the Roman Empire

- 475-1549 AD, Middle Ages

- 1075-1220 AD, The Crusades

- 1225-1274 AD, Thomas Aquinas

- 1375, Wycliffe New Testament written

- 1449-1800 AD, The Age of Discovery

- 1450-1650, The Renaissance

- 1637-1800, The Enlightenment

- 1454-1455, The Gutenberg Bible was first printed

- 1475, The Codex Vaticanus found

- 1483-1546, Martin Luther, a founder of the Protestant Revolution

- 1517-1648, The Protestant Reformation

- 1509-1564, John Calvin, the founder of Calvinism and reformed theology.

- 1525, Tyndale New Testament written

- 1539, The Great Bible written

- 1550, Robert Stephanus published the Greek New Testament with standard numbered verses based on the Byzantine family of manuscripts, using, or similar to Erasmus' fifth edition.

- 1560, The Geneva Bible written

- 1582-1610, The Douay-Rheims Bible written

- 1611, The King James Bible written

- 1643-1648, The Westminster Confession of Faith written

- 1662-1714, Matthew Henry

- 1703-1758     Jonathan Edwards

- 1837-1889 Dwight L. Moody

- 1881-1885, The Revised Version of the Bible was published

- 1942, the New American Standard Bible was published

- 1946-1952, The Revised Standard Version of the Bible was published

- 1948, The Dead Sea Scrolls were discovered, containing over 223 original manuscripts and commentaries, written almost 2000 years before, including representatives from every Old Testament book except Esther, which is referred to. These manuscripts are more than 1000 years older than the oldest previously discovered, and confirm the accuracy over the years of copying. Some of these texts are dated within 25 years of the original manuscripts. (*Secrets of the Dead Sea Scrolls*, Randall Price, Harvest House Publishers, Eugene, OR, 1996.)

- 1898-1963 C.S. Lewis

- 1973-1978, The New International Version of the Bible was published

- 1979-1982, The New King James Version of the Bible was published

- 1912-1984 Francis A. Schaeffer

- 2001, The English Standard Version of the Bible was published

- 1918-2018 Billy Graham

- 1938-2000 James Montgomery Boice

The activities and advancements listed between about 400 and 1549AD, were in what the non-Christian world calls the Dark Ages or Middle Ages. They even use the terms *The Enlightenment*, *The Age of Discovery* and *The Renaissance*, to denote the time (about 1450-1800) that our minds pulled us out of what they consider the "stagnation caused by Christianity."

## What Do These Histories Tell Us?

In these three histories of human thinking, I have attempted to include some significant milestones in the progress of neuroscience, philosophy, and theology. These milestones should demonstrate the progress in the development of the western human mind throughout history as it contemplated the deeper questions of how we think, and the practical results of our thinking as relates to the advancement of civilization.

Before about 1500AD, science and philosophy were considered to have been orders established by God, and man's job was to discover that intent. Consequently, huge advances were made in the understanding of natural reality.

As the histories show, neuroscience was unknown through these earlier periods of scientific advancement, so the antiquated beliefs in Dualism; the *body,* and the *soul* prevailed. The *brain* functions had not been discovered; the *mind* was the same as the heart, and the *Holy Spirit* was not understood.

As can be seen, from about 400 to 300 BC, Socrates, Plato, and Aristotle made it clear that any materialism of secular and some Eastern religions, which negated the influence of God in science and philosophy, had the same lethal flaw as it has today, in that it is a belief in anegative; really therefore no belief at all. All forms of atheism were, and still are today, simplifications of reality, which instead of rationally and objectively analyzing evidence, have a built-in, pre-conceived notion that all reality can be explained by science. Any questions left unanswered, such as the ability we have to think, to reason, to choose... the very characteristics that separate us from animals, are not worthy of answering. We are left with the impression that an attempt at answering the more difficult questions is an intellectually wasted effort.

If that arrogance were transferred to our other areas of society, such as science and education, the incredible advancement of our civilization would cease. Perhaps that is the goal of the materialist!

The materialists of the days of the Enlightenment, or today, believe that the *brain* can control consciousness. They do not attempt to explain consciousness, sensation, thoughts, emotions, desires, beliefs, or free choice. Such a *mind* would behave in a deterministic way based upon the limited laws of matter or nature.

Note that in the scientific and philosophical Dark Ages and Middle Ages, from about 400 BC to 1700 AD, there was practically no advancement in those disciplines, while during those same periods, there was an explosion of theological advancement.

**This period of documentation, reaction, and response to the physical presence of the supernatural in the natural realm for the first time, in the form of Jesus Christ, permanently changed the intellectual processes of civilization.**

Huge advancements in the understanding of spiritual reality occurred without as much thought into scientific or philosophical advancement. All of the intellectual energy of the world was absorbed by newly recognized and appreciated spiritual truths, since that truth seemed to exist at a higher level than the natural truths of previous generations. Or was it the completely revolutionary theological concept that God could actually enter the natural world as a man, and speak first hand of truths only hinted at previously?

After about 1700, scientific curiosity arrogantly insisted that it did not need a god to explain reality; all could be discovered and explained by the human mind.

The start of this odd two thousand year reversal of interests from the natural to the spiritual, and back again, can possibly be explained by the decline and fall of the Roman Empire; by the lack of available funding for science and philosophy; because of political decisions that forced people to just survive; because the growing Roman Catholic Church had a vested interest in promoting things spiritual; but whatever the reason, the thoughts, and passions of the civilized world were redirected for two millennia. This explosion of unique thought burst into the world with the Incarnation of Jesus Christ, with the result that even the dating process of the world was changed.

Humans realized that they had inherited a sin nature from their desire to take God's place. Jesus, as God, provided a perfect sacrifice to God, through His death, in order to pay the penalty of the separation between humans and God that had been caused by this sin nature. This concept was so radical, so repulsive, so anti-religion, that this man Jesus was murdered in His innocence, thus allowing the successful sacrifice. Indeed, it took mankind around 1700 years to digest and accept this radical truth.

Interestingly, the Scientific Revolution, from about 1550 to 1700, pulled much intellectual curiosity away from the spiritual truths and reverted the energy and the interest back to the simpler and more understandable natural truths. But the concept of total reality was never completely lost. In the mid-1600s, Rene Descartes (1596-1650) again proposed the theory of Dualism to relate the brain to the mind. Initially, in 1644, in his *Discourse on the Method and Principle of Philosophy*, he described the *mind* as the faculty of judgment. His investigations into the dualism of the *mind* and *body* were based on both theology and physics and argued for the existence of God. Descartes reasoned that two substances are distinct when each can exist without the other. He combined the *mind* and the *soul*, as one must do if they attempt to simplify our essence.

But it was not until the late 1800s that neuroscience discovered the electrical transmissions from the *brain* through neurons, and the early 1900s before we began to understand the true function of the *brain*.

For the general public, there were major advantages in returning from the spiritual back to the natural realm, including:

- The causes and effects of the natural world are observable to the initiated, and we assume that we can depend on scientists to tell us the truth, even when that "truth" is only a theory.

- Unproven spiritual theories and truths can only be communicated supernaturally, and it is impossible for many people to possess the ability to access those truths.

- Theological, or spiritual truth requires a God, an intelligent designer, or at least a higher power to rationally function. The acceptance of that God is awkward for many people who find it much easier and simpler to just skip that step, and blindly accept what others have said.

- Acceptance of a higher power who is in control of ourselves may seem like a giving up of the sacred right to ourselves, and that may feel like a defeat. But realistically, that is exactly what is required to understand spirituality and receive the key to understanding life, death, eternity, and reality. Any thinking person will refuse to use blinders in order to make their life simpler when those blinders require the belief in a negative and the lazy refusal to look at all possibilities of existence.

Since science did not understand the functions and workings of the *brain* until the second half of the eighteenth century, the weak philosophical theories of the 400's BC prevailed over these two millennia, and influenced the early Christian theologians to accept a limited view of human thinking and mental existence, so that most, including St. Augustine, 350-430AD, and even St, Aquinas, 1225-1274, were trapped inside the limited duality view of mental existence.

So, as can be seen from the Histories above, in the Enlightenment, the natural sciences became more independent and self-absorbed, to the point that they did not understand or appreciate that the explosion of theological intellectual interest was continuing, and even accelerating.

After the Scientific and Philosophical Dark Age and Middle Age Christian advancement, which took over the philosophical and scientific efforts, there has been a revival of science to the point where now it is probably the most respected and, perhaps, accepted area of understanding behind our question of how and why we think. So after the two thousand year growth of theology as the explanation of how we think, the intelligentsia of the world have reverted to the much simpler and acceptable belief that all *thinking* can be explained scientifically, and there is no longer a need for theology.

This is a very efficient solution to the difficulty that theology had been causing the scientific and philosophical worlds for the last two millennia. So, around 1700, a concerted effort began to recover control of the effort to discover how we think. That effort has progressed successfully scientifically, politically, and educationally since then, to the point that all public teaching of spiritual and supernatural thought is defined as a *religion* and prohibited in public education, grants, and subsidies. So the spiritual reality has successfully been banned from secular education, resulting in only partial education and the opportunity of participating in the advancement of the critical questions considered in this book concerning the future of our planet.

It should be no surprise to us that the scientists, philosophers, and educators of this generation have given up what their predecessors considered the basis of their understanding of reality; God's involvement, for the simpler but more self-serving basis of a new understanding of reality without God.

The German philosopher and socialist, Karl Marx, 1818-1883, and the American philosopher, psychologist and educator, John Dewey, 1859-1952, were both influential in promoting this limited secular thinking and education.

The Apostle Paul, a Jewish intellectual and Pharisee, who had spent his life persecuting Christians, and who was later met personally by Jesus Christ, wrote in about 57AD in Romans 1: 18-22:

> *For the wrath of God is revealed from Heaven against all ungodliness and unrighteousness of men, who by their unrighteousness suppress the truth. For what can be known about God is plain to them, because God has shown it to them. For His invisible*

*attributes, namely, His eternal power and divine nature, have been clearly perceived, ever since the Creation of the world, in the things that have been made. So they are without excuse. For, although they knew God, they did not honor Him as God or give thanks to Him, but they became futile in their thinking, and their foolish hearts were darkened. Claiming to be wise, they became fools.*

Ironically, and much to the chagrin of the scientific and philosophy establishments, regardless of scientific, political, and educational efforts, theology and religion have continued to grow in acceptance since 1700, so that today approximately 83 percent of the U.S. population considers itself Christian (*abcnews.go*, Gary Langer, accessed 4/6/19).

But the success of the establishment has virtually removed the consideration of the supernatural realm from its influence on thinking in the scientific, political and educational systems. As a result, civilization has lost the spiritual discernment of the first two millennia and regressed to a very restrictive understanding of reality, which seems to be repeating the intellectual stagnation that science and philosophy experienced earlier following Jesus' birth.

It appears that humankind, in its impulsiveness and lack of patience, demands continuous change; old ideas are antiquated ideas. No one will admit that all truths are eternal. The "limitation" of spiritual truth, as communicated by Jesus Christ, is that it is completely expressed in about 1000 pages of text, written thousands of years ago. *Miraculously*, those texts have survived despite continued efforts to destroy them, in approximately 5500 copies of original manuscripts which are virtually identical. And these manuscripts are more consistent than any other historical writings. If we ignore the Bible as a legitimate document of history, we might as well ignore all history.

When we refuse to consider answers to a question, we are no longer using deductive reasoning or logic. We have instead already reached a conclusion through inductive reasoning, and are only attempting to confirm that conclusion; the very thing that Christians are accused of. In effect, our goal is to prove that we are right in our pre-conceived solutions, as opposed to being proven right, based on the examination of all possible solutions. The result, as we see in our current civilization, is that many people believe that nothing, including thought and truth, can exist without being natural and, therefore, scientifically explainable.

So, now that science is beginning to understand the purpose and function of the *brain*, and its close companion, the *mind*, is it logical and helpful to separate our mental and physical constituents, or combine them into simpler units, which are easier to explain and control? It appears that two of the three areas of our study into how we think; theology and science, are at fault in stymieing our understanding of reality; theology in its never-ending fight with science for the minds and loyalty of civilization; and science, as the perceived *enemy* of theology, in its never-ending paranoidal fight to control the minds of civilization, and protect its own *territory*. Both professions have succumbed to the temptation to speak authoritatively outside of the area of their expertise and experience and ended up with a 2000 year long unresolved fight for recognition and control of the world. Talk about intellectual immaturity.

Where does this godless trend in discovering how we think lead? We obviously retain more accurate documentation of the most recent era of change in society's way of thinking, which enables us to observe shorter cycles of change. For instance, consider the effects of colonialism, Nazism, the Soviet Union, Islam, and the United States. In our area of interest, we could argue that long term trends are becoming shorter. In neuroscience, this is certainly true, but in philosophy and theology, not necessarily. The only increase in the rate of change of thinking seems to be in communication in these areas. In fact, most recent communication seems to be in a study of philosophers and theologians, rather than in philosophy and theology.

Most objective analyses would differentiate between a relatively free society and an oppressive society in their effects on the progress of creative and productive thinking. Therefore we should discount the shorter term failures as anomalies. So, in our concentration on how to save our planet, the long term trend may or may not have begun. The Christian Bible does not predict an evolution of reason and justice in the future; instead, a gradual deterioration of fairness and peace, resulting in an ultimate destruction of the earth, and a replacement with a permanent spiritual Heaven.

## The Importance of Natural Reality

All natural things are organic or inorganic chemically and are the basis, subject and limitation of natural science. To discover more of natural reality, we must use our senses and the senses of others, to observe, learn, and understand. There is a potential for completely grasping the wonders of nature, including the past and present of human existence. If nature was really created by God, we could never duplicate His intricate Creation, but there is still a potential for replacing parts of nature, including human parts, with synthetic parts. The better we understand God's Creation, the more efficient will our efforts in this endeavor become.

In this natural world, things exist, some seemingly permanent, and others temporary, with a beginning and an end. One of the never disproven Laws of Science is that matter can be changed, but it can neither be created nor destroyed. Einstein proved that energy and mass are interchangeable through the Law of Relativity, $E=mc^2$. Therefore, as long as "c", the speed of light is constant, there is a balance in the universe of mass plus energy. If that balance ever has, or ever will be broken, the speed of light must change accordingly.

Given these facts, within the sphere of nature, existence wasn't naturally created, and it can't be naturally destroyed. Within inorganic nature, matter can be changed by chemical reactions. Within organic nature, the change of matter is caused by the separation or combination of carbon, oxygen, hydrogen, and inorganics. Life and existence are dependent on these natural laws created by God, and it is our responsibility to maintain the critical balance of nature, including our environment and our limited natural resources.

Those who desperately fear the religion of the supernatural, have no answers to the bigger questions of life and existence. They have no logic, no reason – only emotion and an uncontrollable anxiety that their individuality is threatened. They have blind faith in their "right to themselves," but no logic or facts to support it. Their faith is complete anti-science, as well as anti-God. A scientist can never say something does not exist; they can't prove that.

Humans were created by God with an innate awareness of God. Some would say "a God-shaped hole that must be filled." Any true intellectual will admit the presence of a higher power, an intelligent designer, someone responsible for the success of our tenuous existence on this planet. What is the answer to the following rather random questions, and many, many more without considering a supernatural reality?

- How did the universe begin? What is the origin of the mass/energy which caused the Big Bang or whatever imaginative theory is proposed?

- Why have astronomers failed to find the location of the Big Bang, even though emissions from that explosion in a vacuum must disperse into a complete spherical expanding volume?

- What is beyond the last accelerating or decelerating particle from the Big Bang?

- Why do some scientists believe that "c," the speed of light, has been constant since the Big Bang, even though energy "E," and mass "m," have obviously changed?

- Why, during the first trip to the moon, did we discover a dust layer less than an inch thick, when evolutionary calculations predicted about 48 inches, causing the lunar module to be designed with huge floats to land on the dust?

- Why are there petrified trees extending vertically through deposited layers supposedly hundreds of thousands of years apart?

- How does prophesy or mental telepathy work?

- Is there a heaven and a hell?

- Is there such a thing as a spirit, a ghost, evil, or even *good*? Why do all civilizations independently define morality so similarly?

- Why does 90 percent of the world's population believe in a supernatural God?

- What is the source of the universal human and animal need for survival and organization? Without such, chaos would prevail because of the Law of Entropy. Remember the book and movie written by William Golding, *Lord of the Flies*? Chaos is the result of natural devolution. Can you throw a book into the air which has been cut up into pieces, containing one letter each, and expect it to land in the order it contained before being cut up, no matter how many tests you run?

- Why has every civilization throughout history prohibited murder, when murder is so convenient to take care of many of our problems? Why has our current civilization gone against that standard in the last few years and encouraged abortion, euthanasia, and suicide?

- We have a capacity to want certain things for ourselves, or for others, which we can imagine, and which are impossible naturally, but spiritually, nothing is impossible. For instance, for someone, or ourselves, to be healed, to resist temptation, to be more kind. It is impossible for a natural person to imagine spiritual things accurately without spiritual input. So, to understand reality completely, we must understand the natural ourselves, and submit the spiritual to a spiritual being, who can communicate with us. Otherwise, we are stuck in a natural world with no hope of discerning its origin, its purpose, its future, or its destination.

- What animal wanted to fly so much that it developed the first feather, then the second after many millennia, just from its own will?

- How did an animal decide, or "nature" decide for it, that it "needed" to have sight? "Wouldn't it be nice to visualize all these things around us which we can only feel? I'll try real hard to see if I can start seeing". "Mother Nature probably wants me to have a fifth sense anyway, so I can be fit enough to survive. Come to think of it, how does nature want anything; how does it design, plan? Is it intelligent? What is the source of its intelligence? Evolution? Evolution from what? Hold it! This logic isn't working. If I could only see, maybe I could work it out."

- How did inorganics evolve into organics, and organics into living organisms?

- How did DNA randomly evolve with a program for life?

Attempts have been made to answer these questions from a natural, scientific base, but failure has been the only result. We gained a lot of theories, but no scientific laws and no rational explanations. That leaves only a supernatural God as a rational answer to these and millions of other questions of life and existence.

## The Natural Permanence of Non-Living Matter

Natural things exist as organics or inorganics. Non-living organics may have an expectancy of limited existence in their current form, but broken down into a simpler form, or in changing form through heat and/or pressure, may exist forever. Many non-living organics are derived from living organics. For instance, plants under pressure can form peat, which under heat and pressure can form coal, which under heat and pressure can form graphite, and possibly diamonds, which can all theoretically last forever. Likewise, non-living organics break down naturally through oxidation (fire or bacterial decomposition) into inorganics, the natural and permanent end to all existence. As per Einstein, mass may be converted to energy through unusual circumstances, but either mass or energy exists forever, given a constant speed of light.

Once the DNA has decomposed in a living organism through death, the remaining organics can combine with inorganics to form soil, sand, and rocks, which theoretically exist forever.

In fact, as far as natural science is concerned, the molten and non-molten rock core of our planet, and of all solid inorganic matter in the universe, is theoretically permanent.

As gases, the elements of oxygen, nitrogen, hydrogen, etc. have existed forever. They can't be destroyed, only changed.

**As liquids, the water cycle has existed forever. There is a mass balance of water in our atmosphere, and all water in one of its three forms has always existed.**

In inorganic chemistry, inorganics can combine with hydrogen to form acids and ammonia, with hydroxide (OH) to form bases, and with each other to form salts, all of which can theoretically last forever.

So inorganics can be changed, but in their most elemental form, exist forever. There is no death with inorganics.

## The Natural Death of Living Matter

Simple organics combined with acids, bases, salts, etc. can theoretically form organic acids like amino acids, proteins, and nucleic acids such as DNA (deoxyribonucleic acid), the building block of life.

Organics can be formed from inorganics or simpler organics into hydrocarbons such as methane ($CH^4$); more complex organics formed from combinations of carbon and hydrogen, such as benzene, ethylbenzene, toluene, and xylene. Even more complex organics formed from combinations of carbon, hydrogen, and oxygen (carbohydrates such as alcohols, ethers, amines, and pesticides), and combinations of hydrocarbons and inorganics, such as insecticides, like nicotine and DDT; and steroids such as cholesterol and testosterone.

The basic difference between living organisms and non-living organics is that living organisms contain this hereditary material called DNA, which stores information as a code made up of adenine, guanine, cytosine, and thymine, ordered into about 3 billion bases (the rungs of the ladder). They are paired up with each other and attached to a sugar molecule and a phosphate molecule (the uprights of the ladder), to form the familiar double helix spiral shape. The order or sequence of these bases determines the information available for building and maintaining an organism. This code programs the living organism to be a plant or an animal, and which type it is to be, including its unique characteristics within that type.

Science has spent billions of dollars trying to create life from the basic elements and has so far, and forever will, fail because DNA is so complex and includes pre-programmed directions to form life, which appear to be supernatural. Science can only create life from life, which contains DNA, in the form of cloning, grown body parts, etc. Nature, not having intelligence, has no ability or capacity to program anything with intelligence.

While inorganic gases, liquids and solids, and even some non-living organics exist forever, why alone in the universe, do living organics have a life span? They don't exist forever. Is it because their DNA is faulty? As explained above, DNA consists of "non-living" inorganic and organic molecules that form genes and stem cells. Stem cells can theoretically reproduce themselves without death, but the organs which they produce, such as the heart and the lungs, are subject to internal and external forces which can cause their death.

What caused this dichotomy in nature? Why, through evolution, didn't inorganics evolve into non-living organics, which evolved into living organics, which were improved in their permanency? Instead, higher levels of evolution have the greatest weakness. They die.

So, theoretically, living organisms should have evolved into permanence, not death. Therefore, there is no rational explanation for death through evolutionary theory. Death must have come from a supernatural source, outside of the universe, outside of our five senses. Death came from God as a curse for our demanding to take His place.

## Do We Accept a Supernatural Reality?

In order to discern the future of species, we must determine whether there is a reality beyond the natural. If so, this supernatural reality must logically control the natural reality, or it could be understood by our natural brains. So, either the supernatural is true, and we don't understand it, or it is untrue, and an imaginary delusion.

Our limitation in the understanding of spiritual reality is that reasoning is a natural ability. Since the spiritual realm overrules the natural realm, the spiritual relationship with the Ruler of the supernatural, God, is our only connection with that realm. And, as the more powerful and controlling realm, the connection is unilateral, from God to us. All we can do is to accept that relationship, to answer the knock, and to submit to the control. That seems easy to say, but so hard to do; to let go of our control and let ourselves be under the control of an external spiritual force we can't see or completely understand. It seems so un-American, so weak, so pathetic. But if we accept the necessarily submissive relationship between the ruled and the ultimate Ruler, and realize that the Ruler is just, fair and desires good for us, we will gladly let go and submit through the belief, faith and new birth that the Ruler has granted us.

Throughout the history of mankind, supernatural and natural reality have interrelated every day, as Satan has attempted to draw us away from God (I Pet. 5:8-9) *"Be sober-minded, be watchful, Your adversary the devil prowls around like a roaring lion, seeking someone to devour. Resist him, firm in your faith..."* and God has continued to prevail. Indeed, every miracle in the Bible, and those continuing even today, demonstrate the power of the supernatural over the natural.

But, a few times in history, the supernatural has intervened in a way that has changed the natural permanently: the creation itself; the writing and preservation of the Old Testament; Jesus' birth, which established the dating system of the world. Jesus' death and resurrection allowed us to communicate with a holy God and live with Him forever; and the writing and preservation of the New Testament.

With that acceptance of the supernatural reality comes the assurance that God will not end our reality with death. He has provided a permanent existence with Himself in Heaven, or in Hell, depending on whether we have answered the knock and given ourselves up in submission to our true father. John 8:51 says, *"Truly, truly, I say to you, if anyone keeps My Word, he will never see death."* Romans 6:23 tells us, *"For the wages of sin is death, but the free gift of God is eternal life in Christ Jesus our Lord."*

Many more verses teach this critical truth, but one important understanding of the issue is Satan's part. Satan, the author of lies, lied to Eve and promised that she would not die if she followed his directions and that she would be like God (Gen. 3:4-5).

Hebrews 2: 14, *"Since therefore the children share in the flesh and blood, He Himself likewise partook of the same things, that through death He might destroy the one who has the power of death, that is the devil."* So, Satan has the power of death; he started it off and changed a created world with no death, to a fallen world with death. He continues today as temptation, lies, and evil in his form as a fallen angel, prowling *around like a roaring lion, seeking someone to devour* (I Pet. 5:8). But Jesus' sacrifice has redeemed those who have accepted Him and defeated death (*...our Savior, Jesus Christ, Who abolished death and brought life and immortality to light through the Gospel* (II Tim. 1:10). So the supernatural reality is permanent and, as Christians, settled forever!

## Is the Supernatural Random?

Plants and lower animals exist, but do they *live*? They are alive, but they do not have the ability to *live* as do humans, in the realization of a spiritual life. Many humans fall into this category; they accept things but have no motivation to change things. They have no mental capacity to understand the greater truth of reality, nor to live within that reality. They just exist with no more impact on reality than a plant or lower animal.

Humans have the opportunity to separate themselves from plants and lower animals because they have a soul, and a highly developed mind open to learning undiscovered truths. Our brain can theoretically be a void with no sensory input. The mind or soul can contain something, but it cannot be a void. No one has the knowledge or the right to tell you that your beliefs are wrong. The truth is that other people do not know. There is a gospel song written by Alfred H. Ackley in 1933, "He Lives (I Serve a Risen Savior)" that says: "You ask me how I know He lives, He lives within my heart."

**Once you recognize the reality of the supernatural, you should logically be curious about its qualities, its origin, its prevalence, its power, and its control. Any reality, whether natural or supernatural, exists because it was designed; it has a purpose. Randomness in the universe and in our thoughts has never been proven to be true.**

Are we guilty of putting human-derived limitations on our belief in God? It seems that even the wisest of us is unable to fathom infinite power, presence, and knowledge. Is God able to read the thoughts of everyone? Control everyone? If not, why not? Who or what limits Him? Who is qualified to judge that God could not or would not do something? His character and His justice are consistent, but He can do anything. He judges His own character, consistency, and justice; we are not His judge.

So if we desire to remove our blinders and seriously try to understand reality and existence, we must examine all possibilities, not just those which will support our presuppositions.

## Beyond our Senses

The supernatural consists of everything unrealized by one of our five senses. Science can neither prove, nor disprove supernatural reality because its practitioners do not have training or experience in supernatural reality. They, therefore, cannot speak of that truth as a scientist, only as a person, no more qualified than anyone else. As an example, science cannot define or explain *love*.

The absence of belief in something does not disprove its existence. Nature abhors a vacuum. It is irrational to try to prove the non-existence of anything; you cannot. It is impossible to document evidence of the absence of something. After all, the next discovery could prove your theory wrong, as has been the case many times in the past in all of the sciences.

So is anything *impossible*? This word is limited to the natural realm. All things are naturally *impossible* in the spiritual realm, but even in the natural realm, a true scientist cannot say that something is impossible, he simply does not know, as stated above. Philosophically, *nothing* cannot belong to a realm anyway, since it does not exist, and is therefore *impossible*. This is the reason for the Law of Conservation of Matter; it is *impossible* to create matter from *nothing*, because *nothing* does not exist, either in the natural or spiritual realms. Our limitation in understanding is limited only in the spiritual realm, where only God understands all.

As discussed above in the section of this chapter titled *What is Reality?* , natural reality is seen to only be experienced through our five natural senses. But what about supernatural reality? We have no supernatural senses...or do we? The trichotomy (three-part) view that Plato and most later philosophers held is introduced in the Bible in both Testaments.

First Thessalonians 5:23 says, *"...may your whole spirit and soul and body be kept blameless at the coming of our Lord Jesus Christ."* (Please see chapter 3 of *A Christian Environmentalist*, by the author, Xulon Press, 2013, for a more complete discussion on this subject). Oswald Chambers in *My Utmost for His Highest* (Dodd, Mead & Company, New York, 1935) says, *"The secret of a Christian is that the supernatural is made natural in him by the grace of God."* In relation to love, Matthew 22:37 says the greatest commandment is this, *"You shall love the Lord your God with all your heart and with all your soul and with all your mind."* So our souls and our spirit are supernatural and separate from our natural bodies, brains, and minds, and our spirits are able, through the Holy Spirit, to understand the supernatural. The power of our spirit is a reflection of God and is in His image. The brain and mind are natural, as explained above, but the soul and spirit are supernatural and determine consciousness and spirituality. Consciousness is a mental state, not a material state. Only our minds can take us beyond the experiences of our senses, and only our minds working in conjunction with our spirit can take us into an understanding of the spiritual reality. Webster's defines *mind* as *The element, or complex of elements in an individual that feels, perceives, thinks, wills and esp., reasons.*

Whether we accept a trichotomy view with an immaterial mind, like Plato; dualism as St. Augustine and St. Thomas Aquinas, with the body containing the brain and the mind and the soul containing the soul and the spirit; or a five-part reality as I, and as I interpret it, the Bible, teaches, it will be logical that the spiritual part of ourselves is linked with a Universal Mind or Spirit. The only reason for being good, for speaking the truth, for showing mercy, and for showing love, is that God is good, truthful, merciful, and is love. So what is our choice concerning goodness and love? Is there good and love in nature, or is the only good and love in God? What about our destiny? Is it death, as nature provides, or is it eternal life, as God provides?

*Chapter 3*

# What is Humanity?

## Introduction

Some of the information presented in this chapter is taken from *The Future of Species*, by this author, Amazon, 2015. In Chapter 2 of this book, we offered a definition and explanation of *existence* and *reality*. We posited that there is a natural reality and a separate supernatural, or spiritual reality and that the spiritual reality created and continues to control the natural reality. As reality's highest form, as established by any objective thought process, humans have a responsibility on our planet to protect and preserve the environment for the sustainability of humanity, and all of the rest of the natural environment. The Bible describes this responsibility as ruling or having dominion over His Creation.

Since humans are obviously the highest classification of animals, what are our limitations and possibilities for preserving our environment? One critical limitation of humanity, as described in Chapter 2, is death.

There has been much discussion recently as to whether humans can exist forever with or without bodies, but with artificial intelligence, containing knowledge from historical mankind. In this case, our brains and/or our bodies would have to be artificial. If our brains were artificial, they would contain no DNA. Therefore, no humanity; no life even.

Another line of thought is that natural evolution will improve the state of humanity so that we can potentially live forever. Is it possible then to defeat death naturally?

The question of existing forever, not as an inorganic, but with intelligence, can be addressed if one accepts the supernatural portion of reality, because *"intelligence,"* whether speaking of the brain or the mind, is limited to the *natural*, and does not involve the *soul* or *spirit.* So any artificially derived being would be incapable of accessing or understanding the spiritual part of reality, and would, therefore, be an animal, rather than a human.And like an animal, could never be able to possess morality, or any fruit of the Spirit as listed in Galatians 5: 22-23, *"love, joy, peace, patience, kindness, goodness, faithfulness, gentleness and self-control."* These are spiritual gifts which would be impossible to program artificially, since they cannot even be understood, or consistently defined, by humans.

The second thought concerning evolution can be answered by recent DNA research mapping, which refutes the classical evolutionary tree. According to this research, evolution between species has no scientific support through DNA analyses. (Nathaniel T. Jeanson, *Darwin vs. Genetics,* Acts & Facts, Sept, 2014; Henry M Morris III, *Your Origins Matter*, Institutes for Creation Research, 2013; Javier Herrero, European Molecular Biology Laboratory, National Geographic, online). There can be no evolution between species, and there never has been. Humans cannot evolve into a different superhuman species with brain or mind properties not now possessed. If we are to improve our natural intelligence, we must learn and store more past and future facts in our brains, and be able to better organize those facts with our minds; we must use more of our brain and mind capacity.

## Our Limited Intellect

It has been said that you can judge the intellectual level of a person by listening to their conversation as an expression of their thoughts. The lowest level of conversation is supposedly about things, a higher level is conversation about people, the higher still, about thoughts, and the highest level is about God. It seems that in the last few generations, strangely enough, in the name of sophistication and education, we have confined our interest to things (natural science), people (sociology), and thoughts (philosophy). We seem to feel that the highest level of conversation and thought (theology) is beneath our sophistication or education. Sadly we are, and teach our children to be, three-quarter thinkers, thereby limiting our minds and our boundaries.

The complex DNA molecules give humans our unique characteristics of intelligence, allow us to use logic, reasoning, and cognitive thinking. But they also give us many common characteristics of lesser animals, such as movement, erect posture, opposable thumbs, and the five senses. This later purpose of DNA is the reason that humans have so much DNA that is also present in the lower animals, especially those with some characteristics similar to humans. Humans have about 3 billion DNA letters containing genes. *Of these letters, about 900 million, or 30 percent are different between humans and chimpanzees.* (Nathaniel T. Jeanson, "Darwin vs Genetics", *Acts & Facts* (September 2014). *The coding in DNA represents the opposite of chance. Randomness in any code sequence destroys the code. ...It is hugely paradoxical that some scientists suggest randomness could have given spontaneous birth to code sequences as super specific as those of the genetic code.* (Michael Pitman, *Adam and Evolution, a Scientific Critique of Neo-Darwinism*, Baker Book House, 1984, p. 53.)

DNA, as presently unfathomable as it is, exists totally in the natural realm and may eventually be completely understood. It programs our brains and our minds, but it does not change, control, or affect our souls or spirits in any way. These parts of humans are strictly supernatural. I believe, both from the Bible and from personal observation, that all humans have a *soul*, which is basically our supernatural character. That *soul* can define right and wrong, selfishness, altruism, justice, and the fruit of the spirit listed above from Galatians.

The *soul* of a human cannot be quantified by science, although the absence of some of these listed characteristics can be painfully obvious in ourselves or others. Christianity teaches these virtues, but they are certainly not exclusive to Christians, or present in all Christians.

The presence of human DNA therefore naturally defines a human and is the reason that murder, including abortion, offensive war, and capital punishment are inhuman and should be prohibited by civilized societies.

I hope and trust that our thinkers, leaders, and teachers will have the foresight, intelligence, and courage to remove their blinders and be considerate enough to release those whom they teach and influence from their control, so that future generations can enjoy the full scope of intellectual discovery and delight through complete freedom of thought and conversation.

## What is Thinking?

Most of us admire people who think, regardless of their education or experience. It is sad and frustrating to see someone who has made up their mind about a subject without being able to explain why. We wonder if they are interested in knowing the truth about a subject, or just have an opinion that seems to conform to the accepted standards of their personal world. A way out of this trap of knowing without knowing why, is to think about all of our beliefs. Very few schools teach thinking; they typically only train students to mimic their teachers. As a result, most of us do not naturally analyze an issue critically. We just base our beliefs about the issue on what we've read or heard. But what if we were required, or required ourselves, to justify every position and belief we hold? The use of thinking, as opposed to memorization, is a way toward that end.

To *think,* according to accepted definitions, is to have a thought in the mind, and to judge, consider, determine, or work out *by reason.* Our mind analyzes memories which the brain has accumulated and stored, and arrives at a thought, which results in another memory, or words, and/or actions. Those memories form from our experiences, which are input into our brains through our five senses, sight, hearing, smell, taste, and touch.

Many of us are missing one or more of these senses, and if the sense has been missing since birth, would have no memories from that senses. What if we had no senses at all? We would then have no memories from any of the five senses. What would that leave for experience? Nothing! We would have no communication with the outside or internal world. All of those sensory experiences were placed by external forces, human or not, accidentally or deliberately, and ended up as a thought to become part of our personality and character. Mostly our parents, but also our teachers, coaches, friends, and family provide this input. So, is our entire character based on chance and deliberate input in our past through the environment?

We have limited our discussion concerning thinking so far, to physical senses and stimuli. But what about non-physical senses and stimuli? And, do any really exist? Can a spiritual force communicate with a person who has no physical senses, or even who does, and affect the processing of their experiences? If so, our personality and character could also be set and/or altered by supernatural forces beyond our, or anyone's natural control.

Do our experiences make a person more intelligent, smarter, or better? What about the types of experiences? Positive, negative, good, bad, just, or unjust? What emotions did they affect? What are emotions anyway? For that matter, what is good, just, fair, etc.?

Is a happy person one who has had more pleasant experiences? Is a good person one who has had more positive experiences? If so, we are truly a product of our environment.

Obviously, inheritance affects our thinking. Not just the teaching of our parents, but the DNA we inherit. It is believed that inheritance affects the way the mind part of our brains processes our experiences into thoughts, words, and actions. It has been proven that brain development can be inherited, even the part of the brain that can prevail. Right-brained people are said to be more artsy, emotional, and to tend toward inductive reasoning, while left-brained people are more mechanical or scientific, and tend toward deductive reasoning. But regardless of inheritance, if there are limited, or no, sensory or spiritually developed thoughts, there perhaps will be no, or limited, personality and character.

So, what keeps a majority of individuals from murder, theft, or other crimes which are illegal in virtually all civilized societies? Is there a moral reason outside of the law that a normal person doesn't murder those with whom they disagree?

I only mention these possibilities to demonstrate that our ability to think completely and clearly appears to be beyond natural control and the only other control possible is supernatural.

Indeed, is *thinking* beyond our inheritance and DNA the very action, perhaps the only action, that a human can take, that transcends the natural/spiritual boundaries? Other than words, internal and external, the only documented crossing of that boundary in the history of the earth, has been the Incarnation of Jesus Christ. No religion, other than Christianity, has claimed such a radical concept as a direct cross-connection by a person exiting the spiritual realm, entering the natural realm, and re-entering the spiritual realm. But, more importantly, it opened a closed door, so that the very sacrifice of that natural/spiritual man Jesus, for all of the sins that the world imposed on itself, allowed those who accept that sacrifice to duplicate that spiritual to natural, and back to spiritual journey that Jesus traveled.

So, it is logical that when one of God's children accepts this free offer of life everlasting, there is also a free offer transferred. It is one of *spiritual thinking*. In fact, God's Word gives us a name for the source of that *spiritual thinking*: the Holy Spirit.

**Yes, the Holy Spirit, God Himself as part of the Trinity, has indwelled the physical, natural bodies of every child of God, and in that indwelt position, allows us to *think* as God does; to be one with God.**

Radical, sure, but when analyzed, is the only rational solution for the future of our planet; to have it replaced by a spiritual Heaven, containing every child of God throughout history, with the everlasting goal of just living with God. If you can imagine a *perfect* "life," what could be better than to live with God? A perfect, eternal life, and one for which we don't have to worry about the details. A life with a perfect God.

One could argue that there is absolutely no evidence that such a thing could be in our future. We can learn to think the way you describe through education and experience – without any spiritual input. Absolutely true… if we require evidence, because evidence is limited in our weak minds to the natural. We can say that we don't believe in the supernatural, only the natural, but there is also no evidence that the supernatural does not exist. You have the right to believe what you want, but your belief makes nothing true. So, a thinking person must keep their minds open to new radical possibilities of truth and reality, or we are forever stagnated, and then there will definitely be an end to our planet.

## What is Critical Thinking?

Practically, critical thinking is being critical toward, or questioning everything you hear or read. It is the recognition that every teacher, author, song, poet, movie, or play has, not only a story to tell, but a belief or agenda to communicate. It is trying to recognize that belief or agenda for what it is and then make a rational decision concerning its validity. It is testing every piece of communication through a personal belief grid to see if it passes the tests of truth and integrity, perhaps even initially disbelieving all communication, regardless of its source, until it is proven valid and correct.

Everyone, regardless of education or experience, has the right, and perhaps the obligation, to question everything, but at least initially, reject nothing. All communication is based on correct or incorrect knowledge, some on wisdom, or the ability to discern and use knowledge to determine truth. But all knowledge is obtained from some natural source, if in the written or spoken form, from a human. Since all humans make mistakes, sooner or later all human communication, unless inspired by God, will contain a mistake, and blind belief will result in a mistaken belief.

All humans think differently. Some think quickly, some more deliberately, and some think as a sponge, just absorbing thoughts from outside and registering them as facts, without questioning. But one who thinks critically uses reasoning and logic before registering those outside thoughts as facts. If they do not pass the critical test, they are registered as beliefs of others *only*.

According to accepted definitions, critical thinking is determining or working out, by careful analysis, conclusions, from known or assumed facts. This definition assumes that in order to critically think to a specific logical conclusion from known or assumed general facts, we must learn to use deductive or inductive reasoning.

*Deductive reasoning* is from the general to the specific, or from a question or a premise, to a logical conclusion. On the other hand, if we have a series of specific facts or beliefs to begin with, we can use *inductive reasoning* to prove these facts and arrive at a general conclusion.

Scientists are taught critical thinking through deductive reasoning. A scientific law is only accepted if it meets three tests; it must have been observed, it must have been documented, and it must be repeatable by peers. Therefore by a series of repeated and repeatable experiments, a theory will be changed into law. Lawyers and investigators use deductive reasoning to determine specific innocence or guilt by examining evidence and motive, and judges and makers of the law use inductive reasoning to develop a general law from specific complaints and defense. For this reason, very few scientific laws are changed, but many laws are changed. General facts seldom change after being observed, documented, and repeated, but specific facts change continuously, and inductively reasoned laws must be updated as truth is discovered.

Concerning Jesus Christ specifically, deductive reasoning will allow us to better reach the specific conclusion concerning the truth of Jesus. We, therefore, need to examine direct observations of Jesus, direct documentation of Jesus, as well as second-hand observations and documentation. But reasoning is a natural quality, and will never result in the acceptance of us by Jesus. That is a unilateral spiritual action that we can receive, but not cause.

## How to Think Spiritually

As previously explained, most ancient and modern thinkers accept that humans are made up of either three parts; *body*, *soul* and *spirit*, or only two parts; *body* and *soul*. The difference is whether there is a separate spirit or Spirit, which is capable of awareness and communication with God.

The ramification of these beliefs is whether the only way to solve a problem, or discern an answer, is physically through natural science, or mentally through philosophy, or whether there is a third way, spiritually through theology. The limitation of thinking spiritually is that it cannot be done through a natural human or an animal. It can only be accomplished by a spiritual being, and the only human access to a spiritual being is the Holy Spirit. That is why Jesus said in Matthew 12: 31.

> *Therefore I tell you, every sin and blasphemy will be forgiven people, but the blasphemy against the Spirit will not be forgiven. And whoever speaks a word against the Son of Man will be forgiven, but whoever speaks against the Holy Spirit will not be forgiven, either in this world or in the age to come.*

The limitation of many liberal thinkers today is that they choose only to use the first two resources, the *body* and the *soul*, to solve problems and obtain answers. Thus, they become forever frustrated with their limited conclusions.

## Thinking with Imagination

The ability to think beyond realistic depiction, is similar, in a way, to the art of science fiction, fairy tales, or works from authors like Tolkien, C.S. Lewis, or J.K. Rowling. This ability seems to be derived from the right or artsy side of the brain.

While I believe that art is a reflection of God's Creation gift to us, I also believe that science is an explanation of God's Creation gift to us. Some of us who are gifted in science also have the ability to depict nature with models (formulae) to help in our understanding. These theories can become scientific laws when observable, recordable, and repeatable. But scientists, like artists, should have the freedom to think beyond the constraints of scientific laws and to use their imagination into theoretical science. The weakness of these journeys, as in art, is that our imagination, if left unbridled, may lead to truth, and may not. Much scientific speculation is taken as fact by the media and population, and later becomes exposed as incorrect, or a downright fraud. So, I would suggest that we accept theoretical science just like we do science fiction andimpressionistic art, with appreciation of imagination, and not necessarily reality.

*Chapter 4*

# Our Body

## Introduction

The next five chapters will describe each of the five areas of human reality with the goal of determining our ability, potential, and limitations in saving the planet. These areas of interest were defined in Chapter 2, under *Natural Reality*, both from Webster's and Biblically.

Let's consider first the body; its capacities and limitations toward our goal. We must admit that we know more about the body than any of the other four parts of our reality. Even with this vast knowledge of our bodies, how often does the physician just not know what is required without more study, more research?

Much of our medical knowledge is limited to recovery of the previous health of the body through chemical medication or physical repair. We have found that the body has an amazing ability for internal automatic repair, but many times a disease or injury does greater damage than our internal systems can resolve.

Two of the more interesting parts of our bodies are DNA and stem cells. The following paragraphs will discuss these organisms and how they affect the integrity of our bodies.

## DNA, Our Genetic Nature

As introduced above, our body, indeed, the body of any organic species, plant or animal, is a *genome*, the total complement of *chromosomes* of a species, including all *genes* and connecting structures. Each *genome* contains all of the information needed to produce and maintain the organism. A *chromosome* is the microscopic bundle of DNA molecules found in the nucleus of living cells which collectively carry the hereditary material called *genes*. A *gene* is the portion of the *chromosome* containing DNA, which is the basic unit of heredity, and which is able to reproduce itself. Some genes make motor proteins which allow us to move through our muscle cells. Other cells which do not move, such as nerve cells, need other types of genes. Most all cells contain a complete copy of our DNA.

The National Human Genome Research Institute estimates that the human genome contains about 20,500 genes along with 100 trillion cells. Each cell contains 46 chromosomes (the bundles of DNA), and 3 billion base pairs of DNA. This means there are about $10^{25}$ pairs of DNA in our body. Interestingly, it has been estimated that there are about $10^{24}$ stars in the sky.

As we grow from a one-cell creature to an adult, DNA *methylation* occurs and increases, which act as "bookmarks", not to change the cell or the DNA, but to help the cell to use the DNA correctly and efficiently through these *epigenetic* marks. During this process, a methyl group is added to the DNA molecule, which promotes or inhibits the expression of certain proteins, depending on which gene it is on, and where on the gene it is located. This *methylation*, once it occurs, is usually permanent. Research has shown that certain early illnesses, accidents, drugs, or other trauma can interfere with *methylation* and cause problems in later life. In this way, the environment is linked to DNA methylation and, therefore, to health.

One of the most fascinating natural parts of living organisms is that they all contain three major macromolecules, DNA, RNA, and protein. In fact, this is one definition of life.

DNA (deoxyribonucleic acid) carries the genetic information in the cell and is capable of self-replication and synthesis of RNA. DNA controls protein synthesis in cells and is the major constituent of the chromosomes within the cell nucleus.

Plant DNA has the same chemical ingredients as animal DNA, but with a different arrangement.

RNA (ribonucleic acid) is involved in protein synthesis and the transmission of genetic information. These complex helix molecules consist of thousands of organic atoms, which contain the programs which cause us to be plants or animals, humans or bacteria, dark or light pigmented, etc. Most of these molecules are inherited and allow scientists to prove ancestral relationships.

Most animals have two sets of chromosomes, one from each parent, while plants can have multiple pairs. An interesting Bible verse concerning the growth provision of DNA programs is Mark 4:28, *"The earth produces crops by itself; first the blade, then the head, then the mature grain in the head."* DNA allows an acorn to grow into an oak tree weighing thousands of pounds, with no input other than solar radiation, water, and soil nutrients.

Nitrogen containing polymer compounds called nucleic acids are present in the cell nuclei. There can be no life without nucleic acids. There are two kinds of nucleic acids; deoxyribonucleic (DNA) and ribonucleic (RNA) acids. These nucleic acid polymers in DNA and RNA consist of three monomers; a phosphate group, a five-carbon sugar, and a nitrogen base. These monomers in DNA are linked to a double helix ladder to the same type with two or three hydrogen bonds, two for adenine and thymine, and three for guanine and cytosine. In RNA, adenine is linked to uracil instead of thymine, and guanine and cytosine are still linked together. (*Chemistry*, Prentice-Hall, Inc., Boston, Massachusetts, 2005, as updated by publications of the National Human Genome Research Institute, NIH.gov).

There is no creditable explanation for how these molecules have evolved from inorganic matter to have the capacity for programming organisms, and why DNA is so similar between species of plants and species of animals, and even between plants and animals. This is becoming more apparent as scientists continue to map the DNA of various species. Therefore, DNA logically appears to have been programmed by a supernatural intelligence which, unlike the god of nature, can plan, make decisions and control His Creation.

Indeed, there is something about *life* that is mysterious. Why has man, with all of his scientific resources, been unable to create life from anything other than life? The answer lies in the design of DNA, which defines life. Anything containing DNA is living, and things which don't are non-living. Science understands the chemical composition of DNA, as explained above, but it does not understand, and cannot replicate, the relationship between the sequence of inorganic and organic non-living chemicals, which compose DNA, nor the programming, inheritance, and characteristics that DNA contains and controls. Science cannot answer the question of why all inorganics never evolved into living organics so they could move about and survive as the fittest. Perhaps because the inorganics were afraid to die! But a world that can reproduce itself is the highest form of existence. On the other hand, it is completely logical that God would create a diverse world containing inorganics that do not die and living organisms that are subsidized by these inorganics. Evolution cannot explain this dichotomy, through chance and randomness, but Creation does.

One of the interesting findings of DNA research is that programmed within these molecules is the ability for an organism to adapt to environmental change. It has become evident that natural selection, as proposed by Darwin, is not contained in nature as evolution teaches, but in the organism itself.

Nature is not a single natural organism which can be broken down into molecules, but is a philosophical category containing matter, organic and inorganic; as a city contains buildings and people, but cannot be tested without quantifying the structures and the population.

## Theoretical Genetics

Charles Darwin knew nothing of genetics or DNA, which is the only real evidence of genealogical relationships. "Because DNA—not a fossil, anatomical structure, or geographical location—is transmitted at the moment of conception, only DNA directly records a species's genealogy" (Nathaniel T. Jeanson, Ph.D., and James J. S. Johnson, JD, ThD, from *Acts & Facts*, published by the Institute for Creation Research, Dallas, Texas, February 2015). Now we have DNA sequencing information from thousands of species.

There are three major lines of genetic evidence that supposedly support evolution:

- Relative Genetic Similarities—"The hierarchical classification of life, based on anatomy and physiology" (ibid, September 2014): At conception, DNA is transmitted through both the sperm and the egg imperfectly. Each generation grows more genetically distant. This theory places the species with similar DNA in the same branches of Darwin's "tree of life" (i.e., man close to apes, farther from reptiles, and even farther from invertebrates like insects). But this categorization allies equally well to proving design.

- Absolute Genetic Differences (ibid, September 2014): The comparison of the expected genetic imperfections in DNA transmission with time can predict differences and similarities. These calculations by evolutionists prove that their theory of predictable DNA change and their timetable of millions of years are radically flawed. For instance about 900,000,000 DNA letter differences exist between humans and chimpanzees, revealing a similarity of 70 percent. Similar DNA is necessary because of the two species' similar characteristics. There are also similar DNA sequences, on a lesser scale, between humans and reptiles and even plants. But according to evolutionary time scale, these human/chimpanzee differences would have occurred in only six million years—much less than predicted. In effect, the classical evolutionary time predictions are being destroyed rapidly as more accurate DNA mapping is developed. Lately there is more evidence of devolution than evolution.

- Junk DNA: "Since the mechanism of evolutionary change is based on genetic mistakes, evolutionists expect the genomes of certain species to be littered with useless (or junk) DNA" (ibid, September 2014). Indeed the very *engine* of evolutionary change is the genetic mistakes themselves, which allow for interspecies evolution. Recent research has shown that more and more of these "mistakes" or pseudogenes, are functional, and there is not nearly as much junk DNA as predicted.

*The coding in DNA represents the opposite of chance. Randomness in any code sequence destroys the code. It is hugely paradoxical that some scientists suggest randomness could have given spontaneous birth to code sequences as super-specific as those of the genetic code* (ibid, September 2014).

If chemicals combined by an electrical force—as many evolutionists claim—to form a compound that could duplicate itself and eventually become a living cell, that compound would have to be a molecule of DNA. In order for DNA to duplicate itself, there would have to be the presence of amino acids, sugar, and phosphate. As further electrical charges produced more DNA, the original molecule would be destroyed, since as a protein, DNA is very sensitive to heat and pH changes. DNA has been produced from a virus but never from an inorganic. A virus can only duplicate itself within a bacterium, plant, or animal cell, under normal conditions. A virus is, therefore, dependent upon higher levels of life for reproduction. Likewise, only the division of a DNA cell can produce DNA, where each original cell unwinds, and against each, a new strand is built up, forming four strands, so that each new double strand of DNA consists of half-new and half-old DNA. (Randy J. Guliuzza, *A Camera Made from Living Tissue! Acts & Facts*, July 2015).

If evolution were true, what is our future? Even though humans have existed with similar DNA for thousands of years (evolutionists claim millions of years), there has been no physical improvement not attributed to diet and medicine. Perhaps our children are right in their prolific imaginations; we will evolve to have superpowers. Isn't flying, super strength, X-Ray vision, and lightning speed preferable? Why has evolution stopped with no evidence over the years of inter-species creatures? Perhaps because God created humans with the abilities and characteristics necessary to survive, and created an environment capable of sustaining that survival. That is a theory that seems much more logical.

There is no mandate from God for us to not use and enjoy the environment for our sustenance. God allows us to kill vegetation and animals for food, clothing and shelter. So the key to sustainable growth seems to be the use and replacement of renewable resources, and the prudent use and substitution for non-renewable resources (see Chapter 15).

*Chapter 5*

# Our Brain

## Introduction

Chapters 5 and 6 will discuss the two parts of our natural body, which enable us to think and act; the *brain* and the *mind*. As explained in Chapter 4, our natural body is the "house" containing our *brain* and *mind*. If the body fails, the *brain* and mind fail with it. The oxygenated blood flow pumped by the heart allows our living *brain* to function. Science has examined our *brain* since the 1700s AD and has some understanding of its function, but it was not until almost 1900 when the electrical function of the brain was discovered. *Webster's* defines *brain* as: *That part of the central nervous system within the cranium that is the organ of thought, memory and emotion.* After all, our *brain* is a physical organ which can be dissected and studied, but it is so complex that it has never been even partially reproduced, even with computer logic. We can live without the operation of parts of our *brain* due to damage, but we are said to be *brain* dead when our heart continues to pump, but our brain waves cease.

The word *brain* is not found in the Bible, since that organ was unknown at the times the books of the Bible were written. The closest comparable word is *heart*. That word has nearly a thousand uses in the Old Testament, consisting of nine different Hebrew words, and sixty-one uses in the New Testament consisting of three different Greek words.

The words translated as *heart* are defined in Strong's (*Strong's Exhaustive Concordance of the Bible, Dugan Publishers, Inc. Gordonsville, TN*), as *feelings, intellect, a breathing creature, observant, the stomach, the center, thoughts, mind and spirit.* There is an inter-use of the English words translated *heart, mind, soul* and *spirit*, because of the weakness of the English language, compared to the Hebrew and Greek.

Some will combine the *brain* and the *mind* into one organ, but for the purpose of this book, they will be separated, since they have different functions, and since different Hebrew and Greek words are used to describe their functions. We will use the term *brain* to denote the natural organ in which is stored the total of all sensory experiences and perceptions throughout our lifetime. The *brain* also contains all the higher centers for various sensory impulses, and it initiates, controls, and coordinates muscular movements. So once the *mind* portion of the *brain* has applied logic and reason to the data stored in the *brain*, an action may be taken using the neurons produced by the *brain* as electrical impulses.

## Left Brained/Right Brained

We speak of left-brained people who are more mechanical and logical, and right-brained people who are more creative and emotional. We even think that certain races or countries tend to produce more left or right-brained people. We know that the left and right sides of our brain control mechanical and creative functions, but we don't understand the reason that we are one or the other, except through heredity and/or the environment.

Neuroscientist Michael S. Gazzaniga, (Michael S. Gazzaniga, *The Split Brain Revisited* (Scientific American, 2002, p. 27) wrote: *"The left brain is dominant for language and speech"* and *"major cognitive activities, such as problem solving."* It is inventive and interpretive. *"The right excels at visual motor tasks, including matching words to pictures, doing spelling and rhyming, and categorizing objects."* It is truthful and literal, but it has much less consciousness than the left side of the brain.

The conclusion of initial brain research, and virtually all serious studies on the subject since then, has been that both sides of the brain are significant in allowing us to be a complete, functioning human. The dominance of one side or the other may affect our personality, our marriage, our choice of a church (or lack thereof), but brain dominance does not make one person superior or more intelligent than another.

Over the past few years, the initial brain research in the field of cognitive neuroscience has been so popularized into these left brain/right brain differentiation, that science, psychology, philosophy, and religion have used and misused the categories to demonstrate the different ways people think — or think they think.

Assuming the accuracy of this research, a caution to all of us is that our left brain — which is usually initially dominant — tends to rationalize our memories. If our memories have reverted to the more honest right side, reality may prevail, but if we tend to retrieve memories from the left side of the brain, these memories may be rationalized and justified to conform to our prior beliefs, and may actually be false memories. Dr. Gazzanniga explains it this way: *These findings all suggest that the interpretive mechanism of the left hemisphere is always hard at work, seeking the meaning of events. It is constantly looking for order and reason, even when there is none — which leads it continually to make mistakes. It tends to over generalize, frequently constructing a potential past as opposed to a true one* (ibid, p.30).

So we may not be purposely lying when we misrepresent what someone said; our brains may just be protecting our beliefs, right or wrong.

*Chapter 6*

# Our Mind

## Introduction

What is our *mind*? Our body, effectively, is a tool of our *brain*, which signals it to function. Our *mind*, according to Webster's, is *the element or complex of elements in an individual that feels, perceives, thinks, wills, and especially reasons*. In this definition, *element* means a constituent part. As introduced above, the *brain* and the *mind* are considered the natural portions of our thinking process in this book, and the *soul* and *spirit* are considered the supernatural portions. In effect, the *soul* can be said to express our character.

## The Natural Mind

*Reason,* concerning the mind, according to the Unified Theory of Psychology, is a "computational theory" of the mind that posits that the nervous system is an information processing system. It works by translating changes in the body and the environment into a language of neural impulses that represent the animal-environment relationship. The mind is considered the flow of information through the nervous system, which is initiated in and processed by the nervous system (*What is the Mind?* Gregg Henriques, *Psychology Today*, December 22, 2011).

So the *brain* is the physical substance, and the *mind* causes the neural impulses flowing through our nerves. It is part of that mysterious system of energy that is "experienced," but not seen because it has no matter. Some scientists consider God, energy, mind, and spirit (GEMS) part of this vast recently discovered system, and point to forces such as gravity and electromagnetism as examples.

Einstein, in 1905, developed the Theory of Relativity and explained that mass and energy are interchangeable. Effectively, matter, or mass, is the lowest vibration of energy, and the only form of energy that we can see. Other forms are unseen expressions of energy. More and more of these non-mass, or very low mass forms of energy as forces and particles, are being discovered or theorized, such as quarks, neutrinos, fractons, photons, gluons, and gravitons. In order to even be discoverable, these forces and particles must act on discovered forces or particles. Otherwise, they would not be *natural,* and therefore literally undiscoverable by science. Therefore, two of the potential forces listed above, God and Spirit, are supernatural by reason and science. More information on these tiny particles is discussed in Chapter 13.

## What is Our Mind Biblically?

Biblically, our *mind* is the faculty connected with intellectual understanding, as opposed to our *spirit* or *soul* which are discussed below. Strong's Concordance (*ibid*) lists thirty-nine uses of Hebrew words translated *mind*, in the Old Testament, and fifty-one in the New Testament. Unlike *soul* and *spirit*, *mind* is the English word translated for seven different Old Testament Hebrew words and seventeen different New Testament Greek words. The Strong's dictionary lists various meanings for these words, including *mind, heart, a breathing creature, wind, blowing, cognition, deep thought, the intellect, and inclination.* So it is apparent that translators have not all discovered the same meaning of all of the original words.

As in the case of *soul*, the word *mind*, is used several times in the New Testament in conjunction with *Spirit, spirit, soul,* and *body*.

## Our Mind

Harry Blamires, a student of C.S. Lewis at Oxford University, and a noted British thinker and author, in his 1963 book, *"The Christian Mind"*, posits that: *There is no longer a Christian mind, because:as a thinking being, the modern Christian has succumbed to secularization.* Blamires is teaching that Christians should return to the *mind* of a Christian in order to influence the social, political, and cultural life of our earth.

In this book, I posit quite differently, that the *mind*, as he says, *is no longer a Christian mind*, but for the opposite reason. I agree that the corporate *mind* of our world is *secular*, but I propose that it is because the *mind* itself is *secular*; not only in accordance with Webster's definition, as stated above, but also through reason and through the words of the Bible.

Examples of the Biblical use of the word *mind* are:

- I Samuel 9: 19-20, *"Samuel answered Saul…I will let you go and will tell you all that is on your mind. As for your donkeys that were lost three days ago, do not set your mind on them."*

- Matthew 22:37, *"And He said to him, 'You shall love the Lord your God with all your heart and with all your soul and with all your mind"* (heart, mind, and soul are three different Greek words).

- I Corinthians 14:14, *"For if I pray in a tongue, my Spirit prays but my mind is unfruitful."*

See also Mark 12:30, Luke 10:27, Revelation 2:23.

Science has not been able to completely understand our *mind*, since like the *soul*, it is not natural or physical and cannot be dissected. Psychology attempts to understand the results of the functioning of our *mind* and psychoanalysts attempt to manipulate our *mind* to improve its functioning or make it more normal.

Dan Siegel in *Mind: A Journey to the Heart of Being Human* says that the *mind* meets the mathematical definition of a complex system in that it can influence things outside of itself, is roughly randomly distributed, and will produce a large and difficult to predict result from a small input. In math, a complex system is self-organizing, flexible, adaptive, coherent, energized, and stable. He says that the foundation of a healthy *mind* is the ability to integrate differentiated ideas together. Without that ability, a *mind* disintegrates into chaos or rigidity. *Scientists say your "mind" isn't confined to your brain, or even your body, (Quartz, Olivia Goldhill, December 24, 2016).*

Even more so than for our brains, science has been unable to accurately determine whether our heredity, our environment, our God, or a combination, controls our *mind*. Our *mind* doesn't seem to be as deep as our *soul* since the *mind* functions as only an organizer of our experiences, in order to form thoughts that can result in actions or words.

So regardless of whether we approach the question from a scientific, a philosophical, or a theological perspective, the conclusion remains that in order to understand reality, including the possibility of saving our planet, we must look at both natural reality, as it becomes discoverable by science, and supernatural reality as it becomes discoverable through God. Otherwise, we restrict our search for the truth to a small part of total reality and will be forever confused in our quest.

## The Limits of the Mind

How far can you stretch your *mind*? Is there any limit? We have heard that we only use a small portion of our brains, but that myth has been countered by scientific evidence (Robynne Boyd, *Do People Only Use 10 Percent of Their Brains?* Scientific American, February 7, 2008). But what about our *minds*?

The context of this book seeks to determine whether our *minds* are capable of saving our planet. As we shall see in Book III, some aspects of the environment, of which our planet consists, are limited to the natural, and other aspects include the spiritual as well as the natural. If, as this chapter suggests, our brain and *mind* are natural, and our *soul* and *spirit* are supernatural, or spiritual, then our *minds* are incapable of affecting the spiritual in any way. In order to have any communication with, and effect on, the spiritual reality, we must use our *soul* or our *spirit*. But is the *mind* the connection between our *brain* and our *soul*? No one seems to know, and it shouldn't matter in the context of this book concerning our effort to save the planet. To answer the parallel question of saving our *souls,* though, the connection with the *Spirit* is critical.

*Chapter 7*

# Our Soul

## Introduction

Webster's states: *our soul is: the material essence, animating principle, or activating cause of an individual life.* Psalms 19:7-8 says, *"The law of the Lord is perfect, reviving the soul; the testimony of the Lord is sure, making wise the simple; the precepts of the Lord are right, rejoicing the heart; the commandment of the Lord is pure, enlightening the eyes."*

## The Biblical Soul

The English word translated *soul* is from two different Hebrew words and only one Greek word, so in this case, the translations seem to be very accurate. According to *Strong's Exhaustive Concordance*, Dugan publishers, Gordonsville, TN, there are 470 uses of *soul, souls,* or *soul's* in the Old Testament and 57 in the New Testament. There are at least 6 uses in the New Testament of *soul* in the same verse as *body, heart,* or *spirit*, and since they are different Greek words, they seem to have separate intents. I Thessalonians 5:23 says, *"Now may the God of peace Himself sanctify you completely, and may your whole spirit and soul and body be kept blameless at the coming of our Lord Jesus Christ."* Hebrews 4:12 says, *"For the Word of God is living and active, sharper than any two edged sword, piercing to the division of soul and of spirit, of joints and of marrow, and discerning the thoughts and intentions of the heart"* (See also: Matt. 10:28, 22:37, Mark 12:30, Luke 10:27).

The Biblical definition from Strong's of *soul* is: *a breathing creature, vitality, breath, or spirit*. This is an interesting derivation since the uniqueness of animals on earth is that they breathe, move independently, and contain blood.

Biblically, *souls* and *bodies* can end their existence on this earth by transfer into Heaven, or Hell. (Matt. 10:28, 16:26, Mark 8:36, 37, Acts 2:27, 31, 3: 23, 5:20). On the contrary, the *spirit* and, obviously, the *Spirit,* cannot go to hell.

Our *soul*, our inner life, emotions, and consciousness, our character, our very essence, can be revived by God's Word. It can make us wise, as opposed to smart, and can rejoice our heart in delight. Our *soul* is a deeper and completely supernatural place compared to our *heart* or *brain*, a place to which the Holy Spirit has access, but science has no access, unlike Webster's "material essence" term.

Psalms 103:1-2 says, *"Bless the Lord, O my soul, and all that is within me, bless his holy name! Bless the Lord, O my soul, and forget none of his benefits."* Psalms 139:14: *"I praise you, for I am fearfully and wonderfully made. Wonderful are your works; my soul knows it very well."*

Our supernatural *soul* has the ability to understand what baffles our *brain*, our *mind,* as well as scientists. Unlike our *brain* and *heart*, our *soul* is eternal (Matt. 10: 28: *"And do not fear those who kill the body but cannot kill the soul. Rather fear him who can destroy both soul and body in Hell"*).

During life, our physical *brain* is accompanied by our physical *body*, just as throughout eternity, our spiritual *soul* will be accompanied by our spiritual *body*. In I Corinthians 15: 44, Paul tells us that *"It is sown a natural body; it is raised a spiritual body. If there is a natural body, there is also a spiritual body."*

*During death, one's soul* can be transferred to Heaven, to live forever in love and peace with God and Jesus, or transferred to Hell, to live forever in torment with Satan. (Ps. 16:10, 86:13, Prov. 23:14, Matt. 10:28, Acts 2:27, 31).

*Chapter 8*

# Our Spirit

## Introduction

The Bible, many times, speaks of our *spirit*. Not always the *Holy Spirit*, but the human *spirit;* an inner spiritual ability to be attuned to God. 1 Corinthians 14:14 tells us: *"For if I pray in a tongue, my spirit prays but my mind is unfruitful."* Likewise, 1 Corinthians 2:11: says: *"For who knows a person's thoughts except the spirit of that person, which is in him?"* 1 Corinthians 5:3-5 elaborates on Paul's relation, through his *spirit,* to the Corinthian Church members:

> *For though absent in body, I am present in spirit; and as if present, I have already pronounced judgment on the one who did such a thing. When you are assembled in the name of the Lord Jesus and my spirit is present, with the power of the Lord Jesus, you are to deliver this man to Satan for the destruction of the flesh, so that his spirit may be saved in the day of the Lord.*

In Romans 8:16, Paul says, *"The Spirit Himself bears witness with our spirit that we are children of God."*

Strong's Concordance lists 36 uses of *Spirit,* and 212 uses of *spirit* in the Old Testament from 3 different Hebrew words. It lists 128 uses of *Spirit* and 105 uses of *spirit* in the New Testament from 2 different Greek words. The words translated *spirit* and *soul* have many fewer Hebrew and Greek words than those translated *mind* and *heart*. There is no capitalization of the word *spirit* in the Hebrew or Greek of the Old or New Testaments, so all capitalization was added by the translators. *Spirit* is defined in Strong's as *wind*, *breath*, and *breeze*.

## The Holy Spirit

One of the justifications for a four or five-part human reality is that, in Christianity, there is no question that the *Holy Spirit* exists and lives within the bodies of Christians as it's temple. Therefore, if the *Spirit* lives within Christians, It is part of the reality of that Christian, just as much as their *body, brain, mind,* and *soul,* and consequently a portion of the thinking process. In accordance with the Bible verses quoted above, the *spirit*, or the *Spirit*, is separated from the *soul,* since the *soul* can go to Hell, but not the *spirit*. So they are both supernatural, but only the *spirit* is spiritual.

# BOOK III: WHAT IS OUR PLANET?

*Chapter 9*

# The Original Quality of the Environment

## Introduction

The environment, by definition is: *The circumstances, objects or conditions by which one is surrounded*, Webster's Ninth New Collegiate Dictionary, Merriam-Webster, Inc., Springfield Massachusetts, 1990.

Therefore, the environment is everything that is "natural" and definable by science as real. According to this definition of the environment, if we are going to save our planet, we must stop or alter the human effect on the *natural* circumstances, objects, and conditions that now exist. Our environment is limited by civilization in two ways; (1) its irreplaceable natural resources are diminishing; and (2) its capacity to absorb excess residuals of civilization while sustaining life is diminishing.

Another definition of environment is: *surroundings, esp. the material and spiritual influences which affect the growth, development and existence of a living being. New Webster's Dictionary and Thesaurus of the English Language*, Lexicon Publications, Inc, Danbury, CT, 1992.

According to this definition, we must stop or alter the human effect on the *natural* and *spiritual* influences which affect humans in order to save our planet. As discussed, we can have nothing to do with the spiritual influences, other than communicating with God through our prayers. The Bible says, and I can testify that our prayers are heard, and if in God's will, are answered. So praying that we can take the actions and that God will take the actions to save our planet, is critical.

Our natural environment was created with a finite mixture of chemicals in the air, the water, and the land. These chemical mixtures were perfect for the preservation of civilization, but as a civilization, we daily change the concentrations of these chemicals as we withdraw them from the land primarily, as natural resources. In that "mined," extracted or withdrawn state, they are virtually impossible to replace in the same form and location as before they were removed. Therefore, these natural resources are lost forever to civilization, resulting in a finitely limited storehouse of raw materials for the future. That saddles our civilization with a "use by date."

**We will eventually run out of raw materials; the end of civilization.**

Secondly, we daily change the concentrations of these chemicals in the air, water, and land of our planet, by returning the residuals of our civilization to these three realms; or worse, to a different realm. In doing this, we increase the ideal concentrations that God provided for survival, especially in the air, but also in the water, to the point that the tenuous and sensitive condition of the air we breathe, and the water we drink, is unsustainable.

According to the Law of Conservation of Mass and Energy, which was basically modified by Einstein, mass and energy can be changed, but neither can be created nor destroyed. This means that there is a mass balance of mass and energy in the universe. On our planet, this balance is subsidized by energy from the sun, which controls the temperature, tides, winds, and weather of our planet. But this safe radiation is limited to sunny days and locations, and its resulting wind and water energy are limited by geography. So, during periods of time without sunlight, wind, or water movement, we must store any energy gained. Currently, this means heat-trapping or battery storage.

Everything that could extend or improve our civilization by the use of electric vehicles, computers, artificial intelligence, and alternate sources of energy,costs us permanent loss through the one-time removal of rare and valuable natural resources from our environment, never to be replaced. Therefore, the faster we evolve technologically, the shorter the remaining future life of our planet. These chips, hardware, batteries, etc. required as components in our workstations, tablets, smartphones, PCs, and laptops, are manufactured from raw materials. These raw materials come from fifty of the ninety naturally occurring elements on earth, and many, like hatmium, are rapidly dwindling in supply. (*What Raw Materials are used to Make Hardware in Computing Devices?* Andrew Wheeler, 9/29/18. *Engineering .com*).

So, in order to have any movement toward saving our planet, if we limit our environment to the natural, as in the first definition; science, and its implementation, engineering, is our only answer.

The only way possible for humankind to preserve or save our environment, according to this definition, is to completely understand the natural environment; (1) how it came into being, (2) how it has changed and evolved, (3) its current existence, (4) how to prevent further deterioration, and (5) how to save it, or cause it to return to the condition it was in before the acts of humans began to cause deterioration. The following sections describe the constituents of the environment from their origin to the present.

## The Air Environment

In the minds of thoughtful people, the purpose of saving our planet is to allow human life to continue. At its simplest level, life requires organic food as an energy source, or electron donor; and either oxygen, nitrite, nitrate, sulfate or carbon dioxide as an electron acceptor. Humans, of course, and the higher animals, use only oxygen as their electron acceptor, while the higher plants use carbon dioxide. The oxygen/carbon dioxide cycle of our planet for the higher living organisms, involves the ingestion of oxygen and the exhalation of carbon dioxide. Conversely, plants ingest carbon dioxide and discharge oxygen.

**Therefore the animal and plant lives are critically interdependent.**

Our troposphere is the closest portion of the atmosphere and reaches to 12 miles (20KM) above the earth's surface. It contains half of the earth's atmosphere and nearly all of its water vapor and dust, which is the reason that clouds are found there. The next layer of our atmosphere is the stratosphere, from 12 to 31miles (52KM) above the earth's surface. Ozone is plentiful there and serves to heat the atmosphere and adsorb harmful radiation from the sun. It is very dry and about one thousand times thinner than air at sea level. This is the layer used by jet airliners. NASA says Earth is the *only planet in the solar system with an atmosphere that can sustain life* (Earth's Atmosphere: Composition, Climate & Weather", Space.com, Feb. 2019).

These are the eleven most common elements in our atmosphere in percent by volume in dry air:

- Nitrogen 78.09

- Oxygen 20.94

- Argon 0.93

- Carbon Dioxide 0.03

- Neon 0.0018

- Helium 0.00052

- Methane 0.00022

- Krypton 0.00010

- Nitrous Oxide 0.00010

- Xenon 0.00008

- Hydrogen 0.00005

In addition, water Vapor ranges from a trace to about 4 percent by volume in the atmosphere (NASA).

## The Water Environment

Water is the oldest constituent of our earthly environment. Biblically, it was created first, as described below:

In Genesis 1:1-2, the Bible states:

> *In the beginning God created the heavens and the earth. And the earth was formless and void* (a waste and emptiness), *and darkness was over the surface* (face) *of the deep; and the Spirit of God was moving over the surface* (face) *of the waters.*

II Peter 3:5: "*...the earth was formed out of water and through water.*" Scientifically, water has been around as long as the earth. Charles Fisherman, author of *"The Big Thirst: The Secret Life and Turbulent Future of Water"* says: *"water molecules are extremely resilient, and it is likely that all water molecules present now were the same water molecules available for billions of years."* (As reported in *Quora*, May 30, 2016).

Dr. Conel Alexander, a scientist at the Carnegie Institute for Science, Washington, DC, says, *"Our findings show that a significant fraction of our solar system's water, the most fundamental ingredient to fostering life, is older than the sun,"* as reported in *The Journal of Science* September 26, 2014, and *Huffington Post*, September 26, 2014.

Water, therefore, is completely recyclable and exists on our planet in a closed system. It is always present in the air, the water, of course; the earth, and in organics, in the same total amount. We therefore don't have to be concerned with running out of water, only the distribution and purity of water. Water has the ability of acting as a solvent for most solids, gases, and other liquids, so it does not naturally remain pure. In its cycle through evaporation, rainfall, freezing, and condensation, water absorbs almost everything.

On the earth, the oceans cover about 70.9 percent of the total surface (*nydc.noaa.gov*), and are the biggest receptor, or sink, for carbon dioxide pollution in the air, so it is critical that they and the streams and rivers that feed them, are protected from our damage.

## The Land Environment

The third receptor of pollutants, the land, throughout history, has been the only receptor of non-vapor particulate pollution in arid areas, and certainly, along with water, the primary receptor in all areas of civilization. Even today, the edges of growing urban areas are used for landfills, and garbage dumps for the by-products of development.

Excrement has always been disposed of in streams, latrines, privies, outhouses, or on the surface of the ground.

Of the three pollutant receptors, the soil provides the only significant biological means of pollutant treatment (water can provide some minor biological treatment), in the form of cellular oxidation. Soil can use bacteria which can convert and combine organics primarily into carbon dioxide, water, hydrogen sulfide, nitrogen forms, and methane. Because of the relatively non-porous nature of soil, the result of soil pollution is normally local in concern. In Karst terrain, which is permeated with fissures and caves, pollutants can potentially be transported regionally, either underground or when they reach surface waters.

The disposal of solid wastes on the ground doesn't guarantee their treatment, especially if these solids are placed into cells, such as in a landfill where they are not associated directly with the soil. Inorganics placed in the soil will normally remain as placed, which constitutes storage rather than treatment. I have excavated a fifty-year-old abandoned landfill and found readable newspapers and whole pieces of vegetables.

Therefore, land or soil and its vegetation is a receptor of pollutant discharges from society. Vegetation, depending on its extent and its characteristics, can uptake liquids through transpiration, primarily through its roots. The pollutants in these liquids are either filtered out or absorbed or transpirated into the stems, branches, trunks, leaves, blades, etc. Some pollutants, such as nitrogen, phosphorous, and trace metals, act as nutrients to the vegetation; carbon dioxide acts as an electron acceptor, but some pollutants are toxic to the plants or accumulate within them, making the plants possibly toxic to animals and humans.

Pollutants which are not absorbed by vegetation may be adsorbed onto particles of soil, either from the air or from stormwater or by directly discharged flow. As explained above, bacteria in the soil can act to break down organic pollutants, ultimately into carbon dioxide and water. Bacteria have no effect on inorganic pollutants, other than valence change, or as anaerobic reducing agents affecting $H_2S$, $NH_4$, $PO_4$, etc. These inorganic pollutants can be considered to "have returned to the soil," although not necessarily in locations and concentrations at which they were initially removed from the soil.

## Our Relationship to the Environment

Each day we arise and follow pre-planned, or spontaneous activities, actions, and reactions. We go to work, shop, exercise, meet with people, read, study, travel and perform many different functions for survival or pleasure. Nature affects and changes these normal activities with rain, heat, cold, catastrophic weather, etc.

Conversely, the human effects on nature can also affect our daily lives, when we become sick from environmental pollution, herbicides or pesticides; cannot afford fresh vegetables because of our urban location; cannot enjoy nature because of human desecration; or when we experience the effects of acid rain, smog or tobacco smoke.

Have you ever felt at one with nature? You can't be at one with nature, only with God. Nature, in its fallen state, can hurt or kill, when you least expect it. But you can protect yourself from nature and, at the same time, protect nature. When you analyze nature, everything about nature is designed to return living organisms to the soil. And this is the best we can expect from evolution? Fortunately, God has endowed us with a brain, a mind, a soul, and a spirit to control nature to a limited extent.

Since each of us is affected by the environment, and also affects the environment, we must make conscious decisions in our relationship to, and with, nature.

We have been given dominion over the environment; to have an influence over and to guide. So our dominion over nature, as over our children, includes guidance. In the following two chapters, we will examine the current state of the environment and the regulatory interests and trends.

*Chapter 10*

# The Residuals Entering Our Environment

## Introduction

The three realms of our natural environment receive un-naturally caused organic and inorganic pollutants from the other realms as well as from the life or growth within the realms. These pollutants are transported from the realm in which they were created by natural processes, or by human activities (which could be defined as natural activities just as well as can animal activities), distributed into a different realm or stored back into the same realm, These pollutants may be identical to their created form or may be chemically altered by natural or human activities or processes. Some examples of such pollution are as follows:

## Natural Pollutants

Even without the existence of humans, nature, including vegetable and animal life, produces a large amount of waste through death and decay, animal excrement, forest fires, volcanoes, earthquakes, tsunamis, etc. Except for volcanoes, a small amount of these residuals end up in the air. The residuals entering the soil are mostly recycled, but the runoff of natural residuals to bodies of water can cause around 25 percent of water pollution.

It has been estimated that domestic animals use 30 percent of the world's ice free land, consume 8 percent of the world's fresh water, and produce 18 percent of the world's greenhouse gases, which is more than all forms of transportation (*Meat from a Petri Dish,* Scott Canon, Kansas City Star, 2/8/19).

Natural decay has been estimated to release over eight times as much $CO_2$ to the atmosphere as the amount emitted by humans. (*U.S. Global Change Research Information Office, Common Questions about Climate Change,* Feb, 2019).

Natural pollutants are those emanating from non-human nature. They have been present in their own environment since the earth was formed. Natural pollutants constitute about half of the sources of pollution of the air, water, and land, so humans are not responsible for all pollution. Some of the natural pollution sources are as follows:

## From Land

- Inorganics such as salts, gypsum, nutrients, and metals

## From Forests

- Carbon dioxide and hydrocarbons from rotting vegetation
- Tannic acids causing lower pH

## From wild animals

- Carbon dioxide from breathing and death
- Methane from flatulence
- Hydrogen sulfide from flatulence
- Ammonia from urine

- Hydrocarbons and organics from death
- Inorganics and organics from erosion due to deforestation by animals

**From Volcanoes**

- Carbon dioxide
- Hydrogen sulfide
- Methane
- Subterranean inorganics
- Inorganics and organics from erosion due to deforestation
- Pollution caused by temperature effects
- Pollution caused by flooding as a result of volcanic activity
- Pollution caused by animal death

**From Earthquakes**

- Pollution caused by death
- Pollution caused by deforestation
- Stagnant water from evaporation

**From Drought**

- Lowered dissolved oxygen in water
- Pollution caused by death

- Pollution caused by deforestation

- Low flow in steams causing lowered dissolved oxygen, and higher water temperatures

**Human Caused Pollutants**

With the passage of time, certain residuals of civilization remain and re-enter the environment through the air, the water, or the land.

From the beginning of civilization, cooking and heating have been the source of air pollution from the burning of animal, vegetable, and petroleum-based hydrocarbons, whether wood, peat, dung, fat, vegetable oil, animal oil, natural gas or coal. In the air, these hydrocarbons (organics), residual inorganic vapors, such as $CO_2$ and $H_2S$, and particulate pollutants receive no beneficial treatment, are not normally purified or broken down, but only dispersed, except for absorption and transfer to the ground by rainwater. Also inorganics entering the air, such as $NO$, $NO_2$ and $SO_2$ can undergo thermal or photochemical reactions in the presence of oxygen and/or sunlight to form different compounds. There also can be reactions on the surface of particulates or in solution that cause some chemical changes.

The cumulative sources of greenhouse gases, mostly $CO^2$, as estimated by EPA (epa.gov, Climate Change Indicators) are as follows:

- Electricity Generation: 31 percent

- Transportation: 26 percent

- Industry: 22 percent

- Agriculture: 9 percent

- Commercial: 7 percent

- Residential: 5 percent

So if we consider our residential use, plus the commercial products we buy, plus the food we consume, plus the industrially manufactured products we purchase, plus our transportation, we are a large contributor to greenhouse gases and potentially to their effect on the air quality of our planet. The land and forests of the US absorb about 11 percent of these greenhouse gases, but the remaining 89 percent finds its way to the country's water and air.

As discussed in Chapter 9, our planet's water is polluted, about 50 percent from agriculture and undeveloped areas, 25 percent from direct discharges such as industries and municipalities, and 25 percent from indirect discharges, such as rainwater runoff from developed properties.

Land pollution consists of municipal, industrial, commercial, institutional, residential, hazardous wastes, and construction and demolition solid waste, placed in landfills for long-term storage. The environmental effect from this storage is from runoff and leachate. The runoff can contaminate the nearby surface waters, and the leachate can contaminate the subsurface groundwater.

Landfill waste can be broken down into the following general categories of major constituents which enter municipal landfills:

- Food-13 percent

- Rubber, leather, textiles-8 percent

- Wood-7 percent

- Paper-31 percent

- Plastics-12 percent

- Metals-8 percent

- Glass-5 percent

- Yard trimmings-13 percent

- Other-3 percent

Of all types of landfills, including municipal, about 21 percent additional volume enters Hazardous Waste or Special Waste Landfills, and about 15 percent goes to Construction and Demolition Landfills.

## Pollution by Agriculture

Like natural residuals, agricultural residuals are additions into the air and soil, and can add another approximate 25 percent of the total water pollution through runoff of soil, fertilizer, herbicides, and pesticides. Agricultural sources of water pollution include:

- Animal origin

- Vegetable origin

- Nutrients, fertilizers

- Pesticides, herbicides

- Rain/storm water

## Pollution by Municipalities

Municipal and industrial residuals make up the remaining 50 percent of water pollution.

Municipalities voluntarily collect and dispose of certain residuals, including sewage, landfill leachate, burned trash and garbage, and stormwater runoff. All of these residual disposals are highly regulated by the EPA, which has improved our air, water, and soil quality since the 1960s, but there are still large amounts of these residuals entering our environment.

Some specific pollution sources from municipal activities are as follows:

- Dwellings
- Commercial development
- Institutions such as schools, health, hospitals, and prisons
- Governmental facilities
- Streets and roads
- Landfills
- Leaking underground facilities
- Rain/stormwater

## Pollution by Industries

Industrial air emissions, water discharges and solid waste disposal are regulated by the EPA. Regulatory improvements continue to be made, but sadly, certain environmental groups have slowed this process by filing lawsuits that tend to delay promulgated EPA Rules and Regulations.

Examples of direct industrial pollutant discharges are as follows:

- Hydrocarbons
- Carbon dioxide
- Inorganics; metals and non-metals

- Volatile organics
- Nuclear wastes
- Transport water
- Washing and rinsing water
- Solubilizing, electroplating and blending water
- Diluting water
- Direct contact cooling or heating water
- Sewage
- Shower and sink water

Indirect sources are non-contact water and include:

- Boiler feed water
- Cooling water
- Heating water
- Cooling condensate
- Transportation within property
- Landfills
- Leaking underground facilities
- Rain/storm water

*The Residuals Entering Our Environment*

The U.S. Greenhouse Gas Emissions Inventory estimated that in 2006, fossil fuel combustion accounted for 94 percent of all human sources of $CO_2$ emissions, 2 percent was from non-energy use of fuels, and the rest was from industrial manufacturing and production.

Of the fossil fuel combustion, approximately 43 percent was from electricity production (83 percent of that from coal), 32 percent from transportation, 15.6 percent from industries, 5 percent from residences, 3.5 percent from commercial developments and 0.9 percent from U.S. Territories). $CO_2$ was 84.8 percent of all human-caused emissions.

## Pollution by Personal Activities

Our personal and family choices have a large effect on the environment. Our personal transportation uses require energy, discharge pollutants and may end up as solid wastes when abandoned.

Our heat for water and homes is provided directly or indirectly by hydraulic, nuclear, fossil fuel, and to a very minor extent, alternative energy.

We discharge our wastewater from urine, feces, garbage disposal, dishwashing, and clothes washing into the ground through septic tanks or to municipal sewage treatment plants. Note that the purpose of the "water" in wastewater is simply to transport the waste to a point of treatment and then discharge it to a body of water.

Since soil biodegrades organics, stores inorganics, and uses water for transpiration of vegetation, if we were able to transfer all of our urine, feces, garbage, dishwashing liquid, and clothes washing liquid properly into the ground, our municipal sewage treatment plants would be unnecessary for domestic purposes.

Personally, our garbage (food wastes) and trash are disposed of privately or publicly through dumps, landfills, or incinerators.

**Dumps and landfills provide only storage, not treatment.**

Human-caused pollutants would obviously be absent in a world without humans. Since they are caused by humans, theoretically, they can be controlled by humans. Roughly half of today's pollution is from human causes, split between agriculture, municipalities and industries; some of which are as follows:

From Breathing

- Carbon dioxide

From Cooking and Heating

- Hydrocarbons
- Carbon dioxide
- Particulates

From Domestic Garbage and Trash

- Hydrocarbons
- Ammonia
- Hydrogen sulfide
- Carbon dioxide
- Inorganic metals
- Inorganic non-metals
- Paper Products
- Food wastes

- Chemical wastes
- Pharmaceuticals
- Endocrine disruptors

From Domestic Animals

- Carbon dioxide from breathing and death
- Methane and hydrogen sulfide from flatulence
- Ammonia from urine
- Hydrocarbons and inorganics from death

All of these human-caused pollutant examples would be more useful if we were to classify them into characteristics that we could use to remove them from the environment. The *Water Quality Handbook, Second Edition*, by the author, McGraw-Hill, New York, 2007, suggests the following classification of pollutants, how they are monitored, sampled, tested and how they pass into the environment:

Physical Pollutants

- Solids content
- Solids type
- Color
- Odor
- Taste
- Conductivity

- Temperature

Chemical Pollutants

- Organic
- Inorganic
- Metals
- Nonmetals
- Acidity
- Alkalinity
- Hardness
- Hydrocarbons
- Fats, oils and grease
- Chemical Oxygen Demand
- Total Organic Carbon
- pH
- Salinity
- Surfactants

Biological Pollutants

- Biochemical Oxygen Demand
- Coliform Bacteria

- Salmonella

- Toxicity

This discussion of the classification of pollutants seems rather complex, but compared to the millions of pages of EPA performance-based requirements, they are minimal, especially as they lead to treatment solutions and not just to enforcement solutions.

Given that the residuals of civilization have always been discharged and must continue to be discharged into the air, the water, and the soil of our planet, just how far are we along with the destruction of our planet?

## Estimating Residual Discharges

Residual discharges may be direct or indirect. Direct discharges are those which are released to the air, the water, or the land either directly from their source, or through a private or public service.

Examples of direct source discharges are all air emissions, including escaping smoke or vapors from fuels, and water discharges through septic tanks, and on-property trash, composting or garbage dumps.

Re-use of residuals would normally be complete (gifts, sales, etc.), but recycling could be only partial. For instance, a scrap iron recycling company typically sends the cloth, leather, and possibly the plastic to a landfill while recycling only the metals.

Examples of indirect discharges are public or private sewerage systems and trash/garbage pick up or hauling to private or public systems.

## Calculations for Direct Discharges

Direct discharges can be estimated as follows:

1. Coal: 5408 pounds $CO_2$ per ton coal plus:

   12 pounds of oxides of nitrogen (NOx) plus

   67 pounds of sulfur dioxide ($SO_2$) plus

   71 pounds of Particulate Matter (PM) (1)

2. Natural Gas: 95.94 pounds of $CO_2$ per 1000 cubic feet of gas (1)

3. Fuel Oil: 18.86 pounds of $CO_2$ per gallon of oil (1)

   (1) Fuel Oil Calculator, USDA, Forestry Product Laboratory, *Natural Gas and the Environment"*, natural gas.org)

4. Wood: 5700 pounds $CO_2$ per ton of wood plus 40 pounds PM per ton of wood ( *Air Pollution from Wood Burning Fireplaces and Stoves,* Toronto Public Health Department, and *The Engineering Tool Box, Combustion Fuels CO2 Emissions*)

5. Septic tank discharges can be estimated at 0.20 pounds of solids per day per person which will eventually need to be hauled, unless it is organic and biologically degraded.

6. Trash and garbage disposal on-site can be estimated in pounds per day as land pollution.

## Calculations for Indirect Discharges

Indirect, or off-site discharges can be estimated as follows:

1. Sewerage system discharges can be estimated at 0.20 pounds of solids per day plus 100 gallons of water per person per day. The weight of the solids in the sewage contains the constituents that can harm the environment. The

liquid, which is virtually all water, is returned after treatment as relatively pure water to the stream downstream from where it was taken.

2. Trash and garbage disposal off-site can be estimated in pounds per day.

3. Durable goods disposed of should be considered trash unless they are recycled or re-used by others.

## Estimating Transportation Discharges

Automobile and truck emissions have been estimated by the EPA as follows:

| Pollutant | Cars, pounds/mile | Light Truck, pounds/mile |
|---|---|---|
| Total Hydrocarbons | 0.006167 | 0.007731 |
| Carbon Monoxide | 0.046035 | 0.061013 |
| Oxides of Nitrogen | 0.003061 | 0.003987 |
| Carbon Dioxide | 0.916 | 1.15 |

(*Emissions Facts; Average Annual Emissions and Fuel Consumption for passenger cars and Light trucks,* EPA, (August 14, 2007)

The emissions from large trucks vary from those of light trucks to about four times those rates. Emissions are slightly more from 10 percent ethanol gasoline.

## Public Transportation

The following are recommendations for $CO_2$ emissions from public transportation:

| Source | Pounds/mile | Description |
| --- | --- | --- |
| Private Auto | 0.96 | 1.58 passengers average |
| Bus | 0.64 | 28 percent Full |
| Heavy Rail | 0.22 | Rapid transit; inter-city, 47 percent Full |
| Light Rail | 0.36 | Electric streetcars; trams, 37 percent Full |
| Commuter Rail | 0.33 | 30 percent Full |
| Van Pool | 0.22 | 56 percent Full |

These are average emissions estimates from *Public Transportation's Role in Responding to Climate Change*, U.S. Department of Transportation, Federal Transit Administration.

They show that normal bus commutes emit a fraction of a normal automobile commute, not considering transportation to and/or from the bus.

## Other Discharges

Individuals may discharge other residuals or have power uses not listed above which will add to the total annual environmental footprint.

## "Green" Solutions Which do Not Solve the Problem

There are many discussed solutions to the problem of residuals entering our environment, which provide us false security at the cost of subsidizing an ineffective industry. The purpose of these examples is to urge the reader to spend time critically analyzing any "environmentally friendly" or "green" product before purchase. Examples of some of these false green products are as follows:

- **Organic Foods from Afar**: The value of organic foods to the environment is the absence of herbicides and pesticides, but if the delivery and purchase transportation cost to the environment overrides that value, these products are no longer green.

- **Cremation**: Huffington Post, states in its web site under environmental costs of cremation, that each cremation uses 28 g of fuel, the equivalent of 563 miles of car emissions at 20 miles per gallon, and emits 540# of $CO^2$ at 2.5 pounds per person per day (see Chapter 13, *$CO^2$ Control*. Those emissions are equivalent to 216 days of average human breathing, or 563 miles of car emissions. Other effects of cremation on the environment are consumed energy (285 Kwh of gas plus 15 Kwh of electricity; as many Kwh as burning a 100W light bulb for 3000 hours), land, building the structure, transportation, and ash containers. For Christians and Jews alike, the Bible mentions no example of a deceased, saved person's incineration. It is, adversely, considered a pagan sacrifice or ceremony.

- **Electric Vehicles**: The World Economic Forum, under battery manufacturing and disposal, states that: "lead acid batteries have a life of about 10 years, and lithium batteries about 2-3 years; about 90 percent of lead acid batteries are now recycled; batteries use nonrenewable hazardous resources such as cadmium, mercury, lead, lithium, nickel and electrolytes; and their environmental affect must include production, transportation, distribution, and natural resources."

Electric vehicle manufacturers publish ranges between battery recharges, but seldom publish the relation between speed and range. I have accessed over 100 graphs and charts showing test results for ranges, and they typically indicate a range reduction of 30 percent to 60 percent comparing 20 and 80 mph speeds, depending on manufacturer, model, and temperature. In addition, temperatures below 20°F will lower range by over 41 percent, and temperatures over 55°F will lower the range by over 17 percent, according to AAA.

Weform.org states that:

*"producing an electric vehicle contributes, on average, twice as much to global warming potential, and uses double the amount of energy as producing a combustible engine car, mainly because of its batteries."*

Much of the excess cost of operating electric vehicles is the production of electricity for regular (150-500 mile) charges. In the US, where 64 percent of electrical production is from fossil fuels, a mid-sized electric car must be driven for 124,000 miles, on average, to break even with a diesel-fuel car, and 59,500 miles for a gas-powered car. Therefore, it takes fifteen years for an electric car to be greener than a diesel car at 8,370 miles per year. (These figures were extrapolated from the report which was based on uses in Germany of only 40 percent coal-fired electricity.) Weform reiterates that battery production causes more environmental damage than car emissions alone. They also state that recycling batteries costs £1.00 or $1.29 US today, but that the reclaimed battery is worth only one-third of that re-manufacturing cost. Lithium batteries cost five times as much to recycle as to make from raw materials. They make the caution that lithium and lithium-ion batteries are flammable and can explode upon disposal.

It is estimated that a battery pack replacement for an electric car will be required at from 100,000 to 200,000 miles at a 2020 cost of about $15,000 plus labor

*Chapter 11*

# The Regulation of Residuals and Environmental Quality

**Introduction**

In air pollution control, under the 1963 Clean Air Act, EPA set certain Ambient Air Quality Standards to protect human health. The EPA has designated that, under the Water Quality Act of 1965, and later under the 1972 Clean Water Act, permits be issued for all direct and indirect dischargers to surface waters of the U.S., based on technology or water quality standards. Solid wastes were regulated under the Solid Waste Disposal Act of 1965, and later under the Resource Conservation and Recovery Act (RCRA) in 1976, and the Comprehensive Environmental Response and Reauthorization Act (CERCLA) of 1986.

Both the EPA, when speaking beyond politics, and the well-meaning environmental control groups, agree that these existing standards have improved the environment since the 1970s, but are not on track to save the planet. Unfortunately, most of the environmental control groups resort to manipulating emotions and legal tricks, rather than turning to scientific data, to make the appeal to the public to reorganize environmental regulations.

Actual graphs produced mostly from government sources are shown in Chapter 1 under *Evidence of Climate Change or Global*

*Warming*. These graphs show evidence of global warming but no evidence of climate change caused by human activities.

## A Rational Method for Pollution Control

We read every day that our civilization is destroying our environment. Before we overreact to that fear, we should objectively examine the effect that civilization has had, and may have in the future, on our environment. An upper-level EPA employee once told me that the biggest problem EPA has with its effectiveness in preserving the environment is the emotional reaction of the media and, especially, popular alarmist books, which often result in complaints, lawsuits, and delayed enforcement. He said that these tactics normally discounted scientific evidence, and damaged the goal of the EPA in their planning for and enforcing of environmental protection.

Along with the return to scientific logic, the EPA must face the fact that it's voluminous and ever-changing listing of environmental regulations, based primarily on performance-based standards, is not adequately protecting the environment, and will never save our planet.

In order to be able to return to a scientific basis for environmental control, we should consider the following:

- **Be more concerned with the characteristics of a pollutant, than its source.** Many source-based regulations are non-uniform and unfair to the environment as applied to different industries at different locations.

- On the contrary, the only variation in domestic sewage quality is typically the local diet habits, and that is a relatively small difference.

- The purpose of using sewers to transfer residuals to a central point for treatment is simply for ease of transportation.

Some residuals are gases, some liquids, and some solids, but we have the technology to change residuals between these three states. Of the states, the liquid state is the easiest and less expensive to use for transportation. So all wastes, gaseous, liquid, or solid must be reused locally or transported to a location set aside for treatment.

**Logically, each generator of residuals, domestic, public, commercial, or industrial, should be responsible for paying the cost of self-treating, reusing, or jointly treating all gaseous, liquid, or solid residuals they produce.**

- If our analysis of this challenge of eliminating the effect of residuals on quality, indicates that some form of joint treatment is preferable, and except for large industries, that should be the case, then a portion of our limited land resources must be set aside for this purpose. That is not necessarily a non-starter. When we realize that if we ask the same questions about other public services, as well as housing, agriculture, natural resources, etc., we arrive at the same conclusion: the ultimate limitation of our land is population, and population growth requires space beyond living quarters.

- Look at exactly how a specific pollutant affects the environment, including human health. This is regardless of whether the pollutant is of domestic, agricultural, or industrial origin. The only type of pollution which can normally be overlooked is natural, and again, nature is capable of self-correction in most natural pollution instances. Effectively, this suggests returning our practices and enforcement to risk-based instead of performance-based standards, a controversy that has haunted the EPA since its inception.

- This effort may involve a different approach for air, water, and land pollutants, since there are different pathways into the environment and into humans. As we have stated previously, the goal of air pollution control should be to return our air to the mixture of chemicals contained before human habitation. Likewise, the goal of water pollution control should be to return our water, including the oceans, to its original constituents. Obviously, water is polluted by runoff from land, but that is natural, and what makes our water rich enough in nutrients to sustain aquatic life. On the other hand, the land is the only realm that provides some treatment to organic pollutants and storage to inorganic pollutants.

- The EPA does not seriously consider the land either a storage resource or a treatment resource in its regulations. It has also treated small service station polluters of the soil more strictly than it has treated large industrial on-site soil polluters. **If every land-disposed pollutant were mapped and categorized, it would at least provide a future resource for extracting these pollutants when their source becomes scarce.** If soil treatment in the form of land farms and waste nutrient soil return for agriculture, were encouraged and regulated, that would give us a disposal resource we don't really take advantage of.

- The Harvard Business School published an *Alumni News* report on February 17, 2019 containing a discussion at the school entitled *Plotting a Path Forward on Climate Change*. One of the participants in this discussion was David Perry, president, CEO and director of Indigo Agriculture. Mr. Perry stated that,

*One third of all carbon emissions are the result of the agricultural industry's use of pesticides, fungicides, insecticides, and chemical fertilizers. Soil from the average cultivated field contains 0.5 to 1 percent carbon, while soil from the average virgin land, including forest and prairie, contains 3 to 7 percent. Carbon capture is the only way to get*

> *out of this. He* further explained that his company commercializes microorganisms that improve crop health and productivity, which reduce the need for chemical fertilizers and pesticides. In addition, he said, *We could absorb all the carbon humans have put into the atmosphere since the beginning of the industrial revolution if we could get [agricultural land] back to that 3 to 7 percent.*

If we have the rationality to return our environmental control back to science, we can finally admit that *yes, we have a problem*, as demonstrated by the facts discussed earlier, in Preface: *Does our Planet need saving?* Our attempt to save the environment without affecting the economy has failed, and we must analyze the cost of actually making progress toward saving our planet. Is our planet worth it? How can the repayment of that cost be accomplished equitably?

So, to face the question of what are the effects of residuals on environmental quality, we must admit that it is possible scientifically to prevent our air from experiencing further deterioration. What is currently in the air unnaturally cannot be practically removed, but all discharges after a reasonable time can potentially be controlled. **We must find out and ask ourselves whether that sacrifice is worth our long term survival, or if we are just willing to pass the burden on to our children.**

We can make similar statements and ask similar questions about the effect of residuals on water quality.

But what about the land? That realm has a certain capacity for waste treatment, unlike the air, which has none, and the water, which has very little. What is the cost of regulating the use of land for treatment, and how far can those regulations go toward minimizing the use of land, or maximizing the efficiency for storage of residuals? Since the total area of land in this country is known and is limited, how long can the storage of residuals last before extraction and reuse of those residuals becomes more feasible than storage?

We are concluding then, that the return of any residuals to the air or water must be prohibited, nor just controlled, and the return of residuals to the land, except for treatment, must be organized and only temporary, if we are to save our planet. Then we must conclude that the effect of residuals on quality is nothing less than deadly.

Are we willing to become the generation to make our planet continually inhabitable?

## Air Management through Regulations

The Clean Air Act in 1963 gave the U.S. Public Health Service, Department of Health, Education and Welfare (HEW), a role in handling air pollution and intervening in instances of health and welfare endangerment.

The Clean Air Act Amendments of 1965 allowed the HEW to set federal automotive emissions standards.

The Air Quality Act of 1967 allowed HEW to issue federal air quality criteria.

The Clean Air Act Amendments of 1970 resulted in the establishment of the Environmental Protection Agency (EPA). The EPA was required to set National Ambient Air Quality Standards primarily to protect health, and secondarily to protect welfare.

The most comprehensive air control legislation to date was contained in the Clean Air Act Amendments of 1990, which addressed pollution from fixed and mobile sources, hazardous air pollutants, acid rain, operating permits, and ozone protection. The Amendments listed six pollutants that were deemed critical in producing harmful effects to public health and welfare. These criteria pollutants were particulate matter, sulfur dioxide, ozone, nitrogen oxides, carbon monoxide, and lead. These pollutants were nationally limited to certain concentrations, and the attainment of these standards continues to affect the local fixed and mobile source emission requirements in the entire country.

The rules restrict development which emits these priority pollutants unless the standards are met in the area. Regulatory authorities in the form of states or metropolitan areas are required to enforce the priority pollutant emissions to assure that the National Ambient Air Quality Standards are met, or to lower the geographical area standards even below the established limits.

Acid rain, caused primarily by the oxides of sulfur and nitrogen from the combustion of fossil fuels, was also addressed by the 1990 Amendments. Mobile sources (automobiles, airplanes, trains, trucks, etc.) are included in these regulations. Other rules apply primarily to industries for the control of hazardous air pollutants. Rules also limit or prohibit certain ozone layer depleting chemicals.

Air pollution control rules continue to be issued and presumably will continue to be issued in the future. New rules apply to specific industries, limiting certain appropriate pollutants to national standards. These standards are based on the Maximum Achievable Control Technology (MACT) experienced in these industries for the control of these pollutants. These regulations are expected to be continually issued in the future as more pollutants of concern are discovered by the EPA, and through risk analysis, health and environmental effects are better understood.

## Air Problems which Remain

According to EPA's current *Air Pollution: Current and Future Challenges*, air problems remaining today include:

- Fine Particulate Matter: *Fine particle pollution... can cause premature death and harmful effects on the cardiovascular system... and respiratory effects, including asthma attacks...*

*Fine particles can be emitted directly or formed from gaseous emissions, including sulfur dioxide or nitrogen oxides.*

- Ozone: *Ozone can increase the frequency of asthma attacks, cause shortness of breath, aggravate lung diseases, and cause permanent damage to lungs through long-term exposure. Ozone is created when emissions of nitrogen oxides and volatile organic compounds react.*

- Climate Change: *EPA determined in 2009 that emissions of carbon dioxide and other long-lived greenhouse gases that build up in the atmosphere endanger the health and welfare of current and future generations by causing climate change and ocean acidification.*

Unfortunately, EPA doesn't comment much further concerning its own conclusions and instead quotes the National Academy of Sciences and *scientific literature* in general, rather than specific studies. The author has specific experience with the interests of EPA as a political tool. This is also obvious with the August 2015 EPA issuance of the Carbon Pollution Standards for power plants, and the EPA proposal of October 2017 to repeal these same Standards. This tendency is also evidenced by both political parties approving EPA Administrators with political rather than technical experience.

- Toxic Pollutants: EPA has listed 187 pollutants in the Clean Air Act *that are known or suspected to cause cancer or other serious health effects, such as reproductive effects or birth defects, or to cause adverse environmental effects.*

- Stratospheric Ozone: *The ozone layer in the atmosphere protects life on earth by filtering out harmful ultraviolet radiation (UV) from the sun. When chlorofluorocarbons (CFCs) and other ozone-degrading chemicals are emitted, they mix with the atmosphere and eventually rise to the stratosphere. There, the chlorine and bromine they contain initiate chemical reactions that destroy ozone.*

## Water Management through Regulations

During the same period that concern was developed over air pollution, water pollution was determined to be a health issue due to the realization that many communicable diseases such as cholera, typhoid, salmonellosis, and amoebic dysentery, are water borne.

As a result of the Clean Water Act of 1972, the EPA required regulatory agencies which, in most cases, were states, to classify all navigable steams for their uses as navigation, recreation, irrigation, water supply, and/or fish and wildlife. The EPA compiled results of a literature review of studies of water quality requirements for each of these stream uses.

*Navigation* pollutants included solids that were floatable or could otherwise interfere with navigation.

Protection of waters for *recreational* uses included pollutants that could be ingested through the skin or orifices.

*Irrigation* water use was protected by restricting the amount of sodium and other pollutants that can negatively affect plant life.

*Water supplies* were protected by limiting organic pollutants which were known or suspected to have negative health effects.

*Fish and Wildlife* is the category of environmental protection for aquatic life in receiving streams. Recommended permit levels were based on a literature review of toxicity test results on standard biological species such as fathead minnows and *sera daphnia*, a water flea. Other permit levels are based on toxicity to the environment from organic and inorganic pollutants, with discharges expressed in concentration. The act was written so that dilution of the concentration with clean water is not allowed (*dilution is not the solution for pollution*). A mass or weight per day limit has been established in most permits for this purpose. Underground water has been classified as surface water for the purpose of permitting.

The first emphasis on discharges was to issue permits for direct discharges to a stream, whether from municipal, industrial, or private sources. These discharges typically enter a stream through one or more pipes or ditches. These sources were thought to include about 25 percent of the total water pollutants being discharged to U.S. streams.

Later, indirect discharges, such as contaminated stormwater carried in storm sewers, or running in sheet flow over the ground, were permitted for municipal, industrial, and private sources. These indirect sources were thought to include another 25 percent of total pollutants.

The remaining 50 percent of discharged pollutants were thought to come from agricultural and undeveloped areas. The only agricultural non-manufacturing discharges typically regulated to date are animal feed lots.

The unregulated agricultural stormwater discharges contain agricultural chemicals such as fertilizer (primarily nitrogen and phosphorous), herbicides, and pesticides. Undeveloped areas, fields, and forests discharge unregulated pollutants into streams and underground reservoirs as by-products of vegetative and animal decay, animal wastes, inorganic metals, and non-metals. **This last category of discharged water pollutants means that unregulated pollutants can be as high as 100 percent of the total pollutant load.**

## Water Problems Which Remain

Mankind has normally disposed of liquid wastes in the water or on the ground. The most common receptor has always been the water, since its usual movement is away from the source of pollution.

The liquid, or even solid pollutants, have historically resulted from eating, cooking, washing, feces, and urine. Where moving streams and rivers are found, civilization develops, and solid wastes such as trash, garbage, and feces have been dumped into these bodies of water for disposal. In the water, the hydrocarbon portion of these pollutants can be broken down by oxygen present in the water into carbon dioxide and water, but much of the time, the bacteria and the oxygen in the water are insufficient to completely break down these hydrocarbons.

Due to the oxidation, along with water movement, tides, or waves, water pollution; even through somewhat diminished natural treatment, is normally only local or regional in scope. But still, organic pollutants can be distributed by the water.

Inorganic pollutants can remain in the water as widely distributed salts or dissolved chemicals, or float or settle in quiescent areas and collect as sediment in the bottom of streams, rivers, or the ocean. Volatile organics, which can evaporate at certain temperatures, and certain volatile inorganics such as carbon monoxide and carbon dioxide, which are disposed of in bodies of water, can return to the air.

## Land Management Through Regulations

The EPA, in 1972, enacted the Resource Conservation and Recovery Act (RCRA), published under 40 CFR 260-40 CFR 267. Under these regulations, any hazardous waste is defined as a solid waste which may:

- Cause, or significantly contribute to, an increase in mortality, or an increase in serious, irreversible, or incapacitating reversible illness, or:

- Pose a substantial present, or potential hazard to human health, or to the environment when improperly treated, stored, transported, disposed of, or otherwise managed.

- EPA established the four characteristics of a hazardous waste as: ignitability, corrosivity, reactivity, and toxicity.

Residuals from industrial waste treatment must be tested to determine if they can be disposed of in a landfill, and if so, what type of landfill.

In 1980, EPA promulgated the Comprehensive Environmental Response, Cleanup and Liability Act (CERCLA), known as *Superfund,* which had the goal of improving past hazardous waste disposal problems. EPA promulgated Title III of CERCLA in 1986, The Superfund Amendments and Reauthorization Act (SARA), which established the Emergency Planning and Community Right to Know, requires industries to inform the public of hazardous substances stored and/or released into the environment.

Discharges into septic tanks or mound systems are usually regulated by counties.

## Land Problems which Remain

In the past, land has been the disposal choice of civilization because of its proximity. Even today, virtually all rural and many suburban houses and commercial establishments use septic tanks and soil disposal for their liquid waste. The reason that solid waste, such as from feces and garbage disposal units is liquid, is so that it can be more easily transported. Ironically, the first step in wastewater treatment is to remove that transport water from the waste. Meanwhile, during transport, the solid waste has contaminated the transport water, so we have two types of waste to treat; solid and liquid.

Other land treatment systems include landfills, as mentioned above, land injection by plowing solids into the surface of soil, land spraying, and using vegetation for transpiration. Other land treatment systems include deep well injection, which is highly regulated and usually prohibited, and incineration, followed by ash disposal on the land.

Another principal source of land contamination is leachate from landfills. When rainwater percolates through a landfill, it can dissolve pollutants and wash them out of the sides or bottom of the landfill. This is especially true with acid rain, which more readily dissolves heavy metals. We completed a multi-year project to measure at a specific location, pH, and sulfuric acid from every rain, frost, or dew with wind direction, using an automatic low volume sampler. The result of this testing was an acid pH as low as 4.0-5.0; 29-43 percent below neutral. **The conclusion, confirmed by EPA, was that the source of the acid rain was large power plants around 100 miles away.**

As mentioned above, sodium in water can blind the soil so that percolation, or the passage of water through soil, is inhibited. This can actually be an advantage in landfill leaching, but a huge disadvantage in agriculture.

One advantage of using the soil, instead of the water or the air for waste disposal is that, ideally, if a potential pollutant is removed from the soil, it should more naturally be returnable to the soil after use. The safest and most long term method of accomplishing this is to return the pollutant in its original form, to the soil from which it was removed originally. This obviously, is difficult and usually impossible.

The alternative is to use the land for *storage* of the used product so that it can be extracted in the future when technology allows.

## Existing Problems with Regulations and Practice

Under the U.S. EPA, there has been disagreement from the beginning between risk-based and technology-based standards. In risk-based standards, the risk to the environment, including human health, is analyzed and regulations developed to minimize that risk to acceptable levels. Part of that controversy has been: what is an acceptable risk? Is one death important enough to allow pollution? The technology-based standards are even less objective, in that they set pollution limits, not on risk, but on the best available, or maximum achievable, technology, regardless of the risk. The government normally sets technology-based standards, although lawsuits and medical and testing advances force the lowering of these standards occasionally.

There are many examples of the weaknesses of unfair technology-based standards: 1) The cost to the economy; 2) The sacrifice of environmental and human health; 3) The cost and difficulty of enforcement. A few specific examples are as follows;

Air Pollution Control: Regulatory requirements and practices force a discharge stack to be taller and make it smaller in order to increase the flow through velocity. This would force the pollutants higher into the atmosphere so that they will not return to the land nearby, at their high discharge concentration, and cause a human health problem. The unintended consequence of these regulations and practices is that the pollutants will eventually settle, become acid rain, or be carried into the upper atmosphere with air currents, causing greenhouse gas blocking of escaping solar radiation, and theoretically global warming. **However, this practice does not prevent pollution; it just dilutes and distributes the pollutants, an illegal practice in water pollution control.**

Water Pollution Control: Domestically, we remove oil, grease, and dirt stains from dishes, clothing, bodies, etc. by saponifying, or solubilizing the pollutant, in other words, dissolving it in water. This simply changes a solid pollutant into a water pollutant. There is indeed an invitation through EPA Regulations to transfer pollutants to a more easily regulated realm, usually water, to simplify control and enforcement, when, as proposed in this book, the logical disposal realm is typically the realm from which the pollutant was removed initially.

We must re-think the entire wastewater transportation and treatment processes to minimize the waste of energy and resources in transferring wastes, such as domestic and commercial wastes, from their natural solid state to a liquid state and then back again. The option is to treat that waste before discharge into highly regulated and environmentally sensitive water, or treating it in the less regulated and less environmentally sensitive land. This change in tradition will cause complaints from both environmental groups, which typically do not understand the complete environmental picture; and also the environmental treatment professions which have sunk huge costs in what could become a futile solution for treatment in the future.

**Now that we are discussing radical solutions to saving our planet, we must decide whether the government has the responsibility to treat all domestic and commercial waste, or whether the waste producers themselves should shoulder that responsibility.** How would a home, for instance, eliminate, or minimize its total waste output to the air, water, and land, rather than depend on the government? Think: reuse, recycle, composting, fertilization, repair instead of replacing, as will be discussed below.

# BOOK IV: OUR LIMITED NATURAL RESOURCES

*Chapter 12*

# The Preservation of Natural Resources

## Introduction

What about the second sensitive part of our planet which affects its longevity – our natural resources? The land of our planet contains virtually all of its non-gaseous elements, and that quantity is fixed. Except for molecular changes caused by pressure, or heat, the elemental and molecular structure of our planet is unchanged with time. If, under the theory of evolution, these elemental quantities just *naturally* evolved from nothing, into the present residuals, that evolution caused a limitation that has pre-set our capacity to extract these elements from the land for beneficial use. Under the theory of creation, God formed our planet out of nothing, because He is God and alone is capable of supernatural/natural transfer. And in either case, once these elements are used, except for the possibility of re-use or recycle, they are forever lost to future generations. So we must conclude that, at some point in the future, all extractable natural resources will be depleted. Beyond that point, we have no further advancement. Our problem is that we do not know when that point will be.

The Age of Computers has excited us so much that some of our more emotional *visionaries* are predicting that our everyday lives, and even our very existence, will be completely dependent on computers in the future. Unfortunately, computers, like any natural product, consist of materials, elements, and compounds which are extracted from the land, and afterward are no longer available to build more computers. The U.S. Geological Survey fact sheet *060-01-USGS Publications*, lists twenty commonly used metals in computers, several of which are rare and/or expensive to retrieve. Other sources list up to fifty total elements. These resources indicate that it takes approximately 1.8 tons of raw materials to make a desktop computer, plus 3.5 pounds of fossil fuel per chip (one or more per computer), plus 400 gallons of water.

On the contrary, one of the wonders of *life* is that it is a renewable resource. What does it take to produce and sustain life? Two animals or plants of different sexes, food, as an energy source or electron donor, and either $O_2$, nitrate, nitrite, sulfate, or $CO_2$, as the electron acceptor. This is radically different from the non-living inorganic portion of our planet, which is limited by definition.

**So once we remove these natural renewable resources, on which we are dependent, we cease to exist.**

What would you rather depend on for your future, a machine limited to the availability of natural resources, or a living and renewable being unlimited by natural resources?

This projection for the end of our planet, as far as I can see, is not even being considered today. If we extrapolate this idea to our original challenge, apparently all *natural* solutions to permanently preserving our planet, are ultimately futile, and the only ultimate solution is supernatural.

Some products used by civilization are renewable. Metals cannot be practically replaced in the same place and condition in the soil after removal, but trees, other plants, which are sources of energy, paper, shelter, furniture, and food, can be regrown and replaced. So the long-term effect on the environment of losing a resource – permanently – is more severe than the short term effect of use and replacement. If more energy and more products were produced from renewable resources, there would be an overall positive effect on the environment.

## Natural Resources used by Humans

Human existence today is sustained by the provision of food, clothing, shelter, transportation, recreation, and electronics, as well as many other necessities. Each of these listed needs will be analyzed as to their sustainability, their effect on the limited supply of resources, and their effect on the environment, all of which affect the future of our planet.

## Natural Resources Used for Food

As introduced above, discounting flying creatures, we have only two sources of food on this planet; the water and the land. There are five essential nutrients for animal survival, according to the US Department of Agriculture:

Water: The most important nutrient; 20 percent typically obtained from foods and 80 percent from water directly. Water also distributes other nutrients to cells to enable vital bodily functions.

Carbohydrates: The main energy source for animals; mostly from whole grains.

Protein: Serves to build and repair tissues such as muscle, skin, bone, and hair. Sources are from lean meat, poultry, seafood, beans, peas, nuts, seeds, eggs, and soy products.

Fats: Serve as back up energy. Sources are olive oil, canola oil, sunflower oil, soybean oil, corn oil, nuts, seeds, avocados, fatty fish and Omega 3 acids.

Vitamins and Minerals: For normal growth and development. Sources are fruits, vegetables, dairy products, and lean protein.

The sources for the four later nutrients are agricultural, animals and sea food. The sources for water are fresh water and desalinated sea water. The agricultural sources require soil rich in nutrients, water, and sometimes chemical protection from disease, insects, and animals. Production is also limited to appropriate growing seasons.

Animal sources for nutrients require the same nutrients as humans along with shelter and appropriate temperature.

Seafood sources for nutrients require pure fresh or seawater, as well as organic food sources.

All of these food sources are renewable, therefore, their sustainability criteria are:

- Available and appropriate land or water
- Necessary nutrients
- Proper temperature
- Proper chemical balance, such as pH

The use of nutrients as food should not negatively affect the supply of resources if care is given to the addition of agricultural nutrients, herbicides, and pesticides, all of which are non-renewable.

Agricultural stormwater runoff is the least regulated of all sources of runoff, which are affected by civilization. It's approximately 25 percent contribution to the pollution of surface water through runoff must be reduced, and existing regulations enforced, in order to protect the freshwater supply.

Other sources of food-related pollution of the water are animal pastures and feed lots, the clearing of trees for agricultural uses, which reduces the $CO_2$ sink, and the runoff from pastures. In areas with dense animal concentrations, methane production is an air pollutant, but the only solution for that issue is fewer animal densities.

Probably the largest source of pollution from food production is in land disposal of waste products. EPA estimates that 22 percent (*"Sustainable Management of Foods"*, EPA) of landfill capacity is taken up by food waste; more than any other single material in everyday trash.

## Natural Resources used for Clothing

Clothing and cloth manufacturing uses agricultural products such as cotton, wool, silk, fur, skins, natural cellulose, and natural rubber, as well as synthetically developed products, such as nylon, rayon, acrylics, polyesters, synthetic cellulose, plastics, spandex, Kevlar, and many others.

The agricultural-based clothing products have a similar effect on the supply of resources, stormwater runoff, water pollution, and limited land, as discussed above, for food production and use.

Synthetic clothing material has a greater effect on limited resources and land pollution since these waste products are less biodegradable. Since agriculture is not used directly for the production of these clothing products, stormwater runoff pollution is less severe than from organic-based clothing or food itself.

## Natural Resources used for Shelter

Shelter needs affect our planet in several ways:

- The use of limited space as a resource.

- The clear-cutting of forests as a resource and as a sink for carbon dioxide emissions.

- The increase in stormwater runoff and associated pollution from developed land, of from 50 to 150 percent.

- The use of renewable versus unrenewable resources for construction and repair.

- The disposal of construction wastes in landfills. In 2015, construction and demolition

    accounted for more than twice the total amount of other municipal landfill wastes. 61 percent of this waste was non-residential; 90 percent 0f the total was demolition. (*Sustainable Management of Construction and Demolition Materials,* EPA).

Unlike food and clothing, shelter needs consist primarily of long-lasting materials such as stone, cement, brick, wood, glass, metals, plastics, synthetic rubber and asphaltic compounds. The use of these materials has virtually no effect on air, unless they are burned; or on water quality, since they are typically inert and non-leachable; but does have a major effect on land because of landfill use for construction and demolition wastes as described above.

Of these raw materials, only wood is renewable.

After construction, buildings require utilities and maintenance to function. The utilities are normally water, sewage, electricity, and natural or LP gas. These are normally provided by a municipal or private utility service. In keeping with the suggested "cradle to grave" approach of determining resource use and environmental effects, these associated utilities and maintenance disposals should be considered.

## Natural Resources used for Transportation

Transportation is one of our controllable actions that affect both the use of non-renewable resources and the deterioration of the air, water, and land. Whether we use public or private transportation, there is still a major cost to resources and the environment. The section *Calculations for Indirect Discharges,* in Chapter 10, recommends methods for estimating various types of transportation costs. The public/ private transportation cost is a factor of cost per mile for purchasing a product, maintaining it, and disposing of it. The "Cradle to Grave" approach described above will allow a more accurate decision and will possibly eliminate some of the "feel good" choices like organic food shipped from afar, synthetic clothing, non-renewable house materials, electric vehicles, etc.

Energy sources used for transportation are gasoline, diesel fuel, jet fuel, electricity, LP gas, and potentially hydrogen, steam, and nuclear fuel. All of these choices are non-renewal except for steam. Steam is an old source of energy which is still used to convert nuclear power to energy, but has not recently been used much with a renewable heat source such as wood. At the present time, hydrogen and nuclear fuel production are impractical on this scale, but they hold potential for the future.

## Natural Resources used for Recreation

Because our use of recreation is minor compared to other needs, the major resource and environmental problems now are confined to the loss of land, which could contain trees that would use carbon dioxide, the stormwater runoff from sports fields, and the transportation to and from sporting events.

## Natural Resources used for Electronics

The EPA publication *Electronics Donation and Recycling* includes a disclaimer, but states that *Electronic products are made from valuable resources and materials, including metals, plastics,* and glass, all of which require energy to mine and manufacture. Donating or recycling consumer electronics conserves our natural resources and avoids air and water pollution, as well as greenhouse gas emissions that are caused by manufacturing virgin materials. For example:

- *Recycling one million laptops saves the energy equivalent to the electricity used by more than 3500 US homes in a year.*

- *For every million cell phones we recycle, 35,000 pounds of copper, 772 pounds of silver, 75 pounds of gold, and 33 pounds of palladium can be recovered.*

Of the more than 50 different elements in a computer, several are rare and expensive. For example, *palladium is now the most valuable of the four major precious metals as an acute shortage has driven prices to a record. A key component in pollution control devices for cars and trucks, the cost of the metal has surged about 80 percent since mid-August* (2019), *making it more expensive than gold.* (The Washington Post, Feb 21, 2019, Why Palladium's Suddenly an especially Precious Metal, by Rubert Rowling Eddie van der Walt, Ranjeetha Puklam, Bloomberg). As our lives become more dependent on computers, we will leave less and less limited resources in the earth to extract, eventually running out of certain raw materials.

## How our Life Style affects the Resources and Environment of the Planet

This Section will examine each of the six needs expressed in the previous Section: food, clothing, shelter, transportation, recreation, and electronics: and make recommendations for modifying our individual and family lifestyles, in order to do our part to save the limited resources and the environment of our planet. The recommendations are intended to be a starting point of resource and environmental awareness, and a motivation for custom tailoring our particular response to lifestyle and needs.

We can do our part of saving the planet as individuals. If enough of us do that, we can really accomplish our goal. Typically, individual activities involve our home and its requirements, including procuring supplies; our job and its requirements; and our spare time activities such as church, recreation, fellowship, outings, sports, etc. The potential discharge into the air environment from these home type activities includes burning wood, coal or gas, or the use of electricity for cooking and heating, burning of trash or rubbish, and emissions from vehicles.

Into the water of our environment, we typically discharge sewage after it is treated in a septic tank and disposal field, sewage after it is treated in a municipal or private sewage disposal plant, and rainwater runoff from the property for which we are responsible.

Onto or into the soil of our environment, we typically discharge our solid residuals such as trash and garbage. Landfills are not disposal systems; they are storage systems. Disposal on land will break down organics only when they are all in contact with soil bacteria, which is virtually impossible to occur in a landfill.

Since soil biodegrades organics, stores inorganics, and uses water for transpiration of vegetation, if we were able to transfer all of our urine, feces, garbage, dishwashing liquid, and clothes-washing residue liquid into the ground, our municipal sewage treatment plants would be unnecessary from a residential perspective.

Our personal transportation uses require energy, discharge pollutants and may end up as solid wastes when abandoned. The current interest in electric transportation may not be an answer, because of the life cycle costs of producing and operating these vehicles, such as battery and vehicle manufacturing and disposal, and battery recharges using fossil fuel energy. This is discussed in detail in Chapter 10.

Our electronics industry is exponentially increasing its use of non-renewable resources. What will happen to civilization when these resources are depleted? We can blindly say that we will develop replaceable raw materials; well, we had better begin that process.

In order to determine how our use of food, clothing, shelter, transportation, recreation, and electronics affects the health of our planet, we need to examine each in two ways. The first is the effect of using a non-renewable resource versus a renewable resource. The second is the effect of the disposal of used materials into the air, water, and land of our planet. The following summary is obviously an oversimplification of a complicated issue, but it is a tool and a call for helping us to be more educated and responsive to a critical need which will affect the future of our planet.

| Need | Effect on Non-Renewable Resources | Effect of Disposal of Waste Products |
|---|---|---|
| **Food** | Major effect from land use for Agriculture instead of forests | Major land disposal effect |
| | | Major stormwater runoff effect |
| **Clothing** | Depends on non-renewable materials | Major land disposal effect |
| **Shelter** | Major effect from all but wood | Major land disposal effect |
| | Major effect from land use fo Shelter instead of forests | Major stormwater runoff effect |
| **Transportation** | Major effect from all fuel sources but wood | Major effect on air from energy wood use |
| | | Minor effect from stormwater |

| | | |
|---|---|---|
| **Recreation** | Minor effect from equipment manufacture | Minor effect on land disposal |
| | | Minor effect on air because of use of property instead of forests |
| | | Minor effect from storm-water runoff |
| | | Minor effect from energy use |
| **Electronics** | Major effect from equipment Manufacture | Minor effect on air from energy use |
| | | Minor effect from land disposal |

We, as individual citizens and employees can have a major effect on the future of our planet if we will do our part, and argue and vote for public policies which will conserve our limited natural resources, and allow the most efficient residual disposal, storage or recycle.

## Our Choices for Food Residuals Disposal

We, as individuals, can reduce the stress on our planet's resources and the environment by implementing the following practices:

- Compost all food waste on-site with the compost product being used for plant food.

- Reduce supplier's transportation resource and environmental effects by buying locally grown fruit, vegetables, dairy products, and meat products.

- Reduce our own transportation effect from purchasing food by buying as close to home or work as possible, or combining trips.

- Purchase organically grown food to minimize herbicide and pesticide runoff and consumption contamination.

- There is no indication that vegetative food is less polluting or resource destroying than animal food.

- Minimize food packaging waste by recycling.

- Raise and grow as much food as possible.

## Our Choices for Clothing Residuals Disposal

- Purchase organic biodegradable clothing such as cotton, wool, fur, leather, silk, etc. There have been complaints concerning the wearing of fur and leather, but those complaints are based on valuing animals more than plants, an irrational belief considering the similar DNA, and the fact that both are renewable.

- Keep our clothing longer and accept traditional styles, rather than mass-market continued replacement styles.

- Recycle used clothing for the needy.

- Recycle packaging.

## Our Choices for Shelter Residuals Disposal

- Use as much renewable material such as wood, as possible for new construction, repair, and in the decisions made for any home purchase.

- Use local construction materials that can be recycled if needed.

- Reuse our houses rather than demolishing and replacing them in order to minimize our residuals.

- Maximize green space for carbon dioxide sinks, and to minimize runoff pollution

- Recycle waste building materials. Wood waste may be useable for heat or power generation.

## Our Choices for Transportation Residuals Disposal

- Minimize the length of trips

- Combine trips.

- Use vehicles with higher fuel mileage rating when possible.

- Keep and repair our vehicles to minimize waste. Objective analyses indicate that the life cycle cost of an automobile is less when repaired than replaced.

- Use public transportation or van pools.

- Walk or use bicycles.

- Use vehicles, if possible, with the following fuels, in order of best to worse in life cycle cost as resources and damage to the environment:

    1. Steam from renewable energy

    2. LP Gas

    3. Diesel

    4. Gasoline

    5. Electrical

    6. Hydrogen (when available)

    7. Nuclear (when available)

## Our Choices for Recreation Residuals Disposal

Normally, our only choice for personal recreation residuals disposal is equipment, which can normally be reused for the needy. We can make our voices heard for manufacturers and sports field managers, especially if we are a major user of equipment or land.

## Our Choices for Electronics Residuals Disposal

Most of the plastics and metals used in electronics manufacturing are potentially recyclable today, and some are routinely recycled. But this industry has not caught up with the metals foundry industry, which was one of the pioneers in recycling, probably because of the profitability.

## Summary of Our Choices

In determining the possibility of saving the environment by using the scientific mind, we must force our logic into the basic constituents of the environment, the capacity of the human mind, and the potential for combining these resources into a solution which can potentially save the planet. That solution must involve an increase in the flow of communication between our research and our thinking population through education. The more experience a person's mind has, the more wisdom it has. The older the average thinking person is, the more knowledge, and therefore the more wisdom humanity has available. That is undoubtedly one of the reasons God allowed the multi-centurions described in the Old Testament. They were necessary at that time to jump-start humanity.

**Einstein said that if we, especially scientists, do not recognize and incorporate the reality of a Creator in our thinking, we are fools.** A thinker wearing blinders is not a thinker at all. As my grandson, Otis, told me the other day, *The only cure for "cloister phobia" is to think outside the box.* So we must think critically, in depth, scientifically, and broadly.

When we reach the point of inadequate natural resources to supply the needs of civilization, including computer power, we must develop substitute materials to extend our planet life. We could also conceive of a more efficient method of storing and retrieving data than computers, but any data management system will require energy and natural resources. Chapter 13 presents some ideas for this possibility.

As we approach that point of no more raw materials, there will be intense competition for the areas of the world containing the remaining resources, as there was last century for oil. Nations will become involved to protect their interests, and international wars will tax our remaining resources. Is this Armageddon? Is this the end of civilization? Possibly, but read on to see what could happen if we accept the supernatural reality as the only long term solution to saving our planet. This means returning science, philosophy, theology and education to its states before the excitement of the potential of controlling our own future through the natural world proved futile.

Ironically, there was a great advancement in visual arts and music beginning in the Renaissance, especially as contrasted with the 20th and 21st centuries, when those forms of art were mostly sacrificed and abused in the names of relevance and diversity, rather than beauty. It appears that science, in its independence of God, is falling into the same trap, and will inherit a similar history of pity and failure, a second Dark Age.

# BOOK V:
# HOW WE CAN SAVE OUR PLANET

*Chapter 13*

# How We as a Society Can Save Our Planet

**Introduction**

All of the residuals of civilization are discharged to the air, the water or to the land after transportation and treatment under the regulatory supervision of the EPA, or its equivalent in other countries. Many of our personal choices can affect the weight and volume of these residuals which enter the environment.

A suggestion for each of us, as individuals or as part of industry, commerce or government, is to examine the extent of the residuals that we can control in terms of pounds per year. Then we can determine if, and how these residuals re-enter the environment.

The suggestions of this chapter are certainly not all conclusive, but instead, are intended to present ideas and plant seeds to begin thinking about a concerted effort to save our planet. How much time do we have? What will it cost and how can it be financed? Some points in this list are in the form of comments and some suggestions, but all are intended to form a possible basis for proceeding with our maximum capability to save, or at the minimum, prolong the life of our planet.

## The Preservation of Humanity

In the recent past, when people were panicked that the expanding population of the world would override our natural resources and available space, certain countries, like China, restricted birth rates. Others, like the U.S., legalized abortion, and the population of much of the world responded by having fewer births. As a result of this global concern of overpopulation, recent generations have less interest and/or belief in the preservation of humanity, as reflected by an increase in abortion, suicide, euthanasia, terrorism, and murder.

This lack of respect for humanity has a dire consequence, which could easily return us to the middle ages. If we analyze the sum of knowledge on our planet, we must conclude that it equals all documentation, plus all knowledge as yet undocumented, which is stored in the minds of living humanity. That means that for every death, we lose permanent access to that knowledge. This lost knowledge could be worthless, but it could help save our planet. In fact, what else could save our planet that is not based on knowledge? This critical, and so often lost knowledge is not confined to the government and the academia; the huge agricultural culture of our planet probably understands more about the environment than all of the government and academic professionals in the world.

As the rate of births in the United States drops rapidly, fewer brains exist each year to accumulate and extrapolate knowledge. If we were able to replace those brains with more capable brains through births, would we increase or decrease our potential knowledge? Since our U.S. educational system is rapidly deteriorating, it appears that, along with a reduced birth rate, we are losing total knowledge. What if we could develop a storage and retrieval system that would allow access to all documented knowledge plus all personally stored knowledge? This appears impossible, but we will never even approach this possibility without trying.

## The Biblical View of the Preservation of Humanity

The Bible, in Genesis 1:27-28, teaches that: *God created man in His image... male and female...and said to them "Be fruitful and multiply, and fill the earth, and subdue it, and rule over the fish of the sea and over the birds of the sky, and over every living thing that moves on the earth."*

Therefore, God created *life*, the *first natural human birth*. Genesis goes on to indicate that there was originally no plan for *death*; His creation of man was: *very good* (Genesis 1:31).

But, I Corinthians 15:21 says: *For since by a man came death, by a man also came the resurrection of the dead.* This explains Genesis 3; the entrance of *death* into a very good world that God created without *death*. So Satan created *death*, the first *natural death*. God responded to humanity's sin in Genesis 3:17-19 with a curse on nature, including humans, and an exile from the paradise of Eden and *eternal life*, (Gen. 3:22-24), but provided for His children to defeat death, through the sacrifice of the man Jesus, which resulted in the resurrection of the natural into the spiritual.

Therefore, because of Satan, we were all born with a curse, an inheritance of sin, and had no means of communicating with a Holy God. Then God, because of His love, sent Himself, as Jesus, into the natural world with the expressed purpose of paying the ultimate debt, through His own sacrifice of being murdered for the sins of the whole created world (John 3:16). That free gift had only one condition: that the saved believe in Him, be born again, the *first spiritual birth*. The promise behind that gift was that *they would not perish*. So, Jesus defeated Satan's *spiritual death* once and for all for His believers. That sacrifice left His believers with *eternal spiritual life*, as was originally intended by God, and left those who refuse Jesus' gift, to inherit their god Satan's gift of *eternal spiritual torment* in Hell with no possibility of relief through death.

I Corinthians 15:53-57 says:

> *For this perishable must put on the imperishable, and this mortal must put on immortality. But when this perishable will have put on the imperishable, and this mortal will have put on immortality, then will come about the saying that is written: "Death is swallowed up in victory. O Death, where is your victory? O Death, where is your sting?"*

The sting of death is sin, and the power of sin is the Law; but thanks be to God, Who gives us victory through the Lord Jesus Christ.

Jesus promised in John 5:24, *"Truly, truly, I say to you, he who hears My Word, and believes Him Who sent Me, has eternal life, And does not come into judgment, but has passed out of death into life."* Again, in John 8:51, *"Truly, truly, I say to you, if anyone keeps My Word he shall never see death."*

James 1:15 elaborates on the relationship between sin and *death* through the Fall, *"Then when lust has conceived, it gives birth to sin; and when sin is accomplished, it brings forth death."*

In summary, the biblical teaching concerning the preservation of humanity is to accept that we have an inherited a sin nature; to accept that God has given us the free gift of forgiveness for that sin nature and the sins resulting from it, and thereby, cover that sin nature in ever increasing obedience. In that obedience, preserve our bodies as temples of the Holy Spirit, God in us, and fulfill our Biblical obligation of having dominion, or rule, over the environment to preserve it for the furtherance of humanity.

## The Secular Effort to Preserve Humanity

As has always happened in the past, science is responsive to public concerns. Interestingly, most of the revolutionary scientific advances recently, which address the preservation of humanity, involve the use or reuse of God created living organic matter.

Science is no longer seriously trying to create life from inorganics or even non-living organics. Instead, science has discovered that genes, DNA, and stem cells can only be productively altered and used from existing genes, DNA, and stem cells. God's original Creation must be a part of the progress or we will be forever frustrated with our inability to advance scientific knowledge.

So, looking ahead with this goal of saving the planet, we must first provide more efficient and longer lasting medical repair and replacement technologies so that humans can live longer, more productive lives, and share knowledge.

The current belief, bolstered by technology, of many younger people, that wisdom and subsequent advancement can only come from the young, has been proven to be indefensible. Knowledge is accumulated in our brain, as explained in Chapters 2 and 4, through sensory perceptions, and the more of these perceptions accumulating in all healthy brains with healthy minds, the more wisdom and technical advancement we as a human race can realize. We must use the talents of the young in technological communication, and combine them with the wisdom of the more experienced in natural science, to advance the cause of saving our planet.

Along with the effort of increasing the knowledge base of humanity, must come a concerted effort to preserve the most important level of existence; animal life, especially human life.

So the methods we have available to preserve humanity include the following:

- Encourage marriage and procreation

- Prevent avoidable deaths such as from tobacco, accidents, poor eating habits, lack of exercise, and lack of access to medical care.

- Prevent avoidable brain cell damage from marijuana and other drugs. (*The Effects of Marijuana on your Memory*, Harvard Health Publishing, accessed 12/20/19)

- Prevent murder, including abortion, euthanasia, and suicide.

- Encourage research into medical and physical practices to use the living organisms God created, such as stem cells and DNA to allow longer life.

As an example of the future of this effort to save humanity along with its collective knowledge, let's examine some of the newer areas of research and progress.

## Stem Cell-Based Therapies

Some of the information in this section is taken from *Stem Cell Information*, National Institutes of Health, www.nih.gov. We have seen in Chapter 1 that matter consists of inorganics, non-living organics and living organics; each of which is more complex than the simpler building blocks. Living organics are differentiated from the others through the presence of DNA. Within the living organic category of matter, there are plants and animals, each of which contain DNA in their stem cells. These cells are ageless, are capable of dividing and renewing themselves throughout the organism's life, are unspecialized, but in animals can form specialized cells, such as muscle cells, red blood cells and even brain cells. In plants, stem cells likewise never undergo an aging process, can self renew, and can form new specialized and unspecialized cells, which constantly form structures and organs such as leaves and flowers. Plant stem cells provide protection to plants to withstand environmental stress since, unlike animals, plants are immobile.

Internally signals are controlled by a stem cell's genes, located across the strands of DNA, which carry coded instructions for all cellular structures and functions.

In animals, embryonic unspecialized stem cells spontaneously form specialized stem cells when they clump together and then form any type of cell structure, such as the heart, bones, nerves, lungs, skin, sperm, eggs, and other tissues. Adult stem cells can also be unspecialized or specialized and serve to generate replacements for cells that are lost through normal wear and tear, injury, or disease, but so far, they are more limited in the various types of cells they can form, than embryonic stem cells. They are less likely to be rejected since they typically are from the person being treated. Stem cells have the ability to replicate themselves, so they theoretically can live forever, like plant stem cells, as long as the body is alive.

Much research is now concentrated on the programming and reprogramming of adult stem cells to function as replacement or renewal for various parts of the body. Induced pluripotent stem cells are adult stem cells that have been genetically reprogrammed or de-specialized to an embryonic stem cell-like state.

## Immunotherapy

One type of medical therapy which has attracted much attention recently is immunotherapy, which aims to reinforce our own body's existing arsenal against foreign bodies and harmful cells, such as cancer cells. Dendritic vaccines are one type of immunotherapy in which dendritic cells are collected from one's body and armed with tumor-specific antigens, and injected back into the body to boost the immune system.

This information on immunotherapy was taken from *The State of Cancer: Are we Close to a Cure?* Maria Cohut, Medical News Today, March 2, 2018.

## Therapeutic Viruses

Another advance in the fight against cancer is the development and use of therapeutic viruses, which can attack certain cancer cells while leaving healthy cells alone.

## Cancer Tumor Starvation

Tumors, as any living organism, require certain nutrients, amino acids, enzymes, and vitamins to grow. Some aggressive cancer growths have been halted by starvation.

## Microsurgeries

Continuum micro robots have achieved motion resolution of one micron or less; 20 percent of the size of a human blood cell.These are being tested for surgical intervention and cellular imaging in invasive surgery. (*Robot Prototype shows Promise for Microsurgeries on Eyes and Aneurysms, Solutions*, Vanderbilt School of Engineering, 2019-2020).

## Nanotechnology

Microscopic nanoparticles have been developed. They can target cancer cells without harming surrounding healthy cells. Some of these therapies have incorporated high temperatures to shrink cancer tumors. Other uses of nanoparticles are to transfer cancer drugs to specific cancer stem cells. Nanoprobes have been used to find specific cancer cells and tumors.

## 3D Printing

It was reported in *Advanced Science,* April 15, 2019, that a team of Israeli scientists 3D printed a heart using a patient's own fatty tissues, and then separated and reprogrammed the cellular and acellular materials. Stem cells were then created that became heart cells. Because the prototype heart was made from a patient's own cellular material, rejection was minimized. 3D, or Additive Manufacturing Technology, has drastically reduced the time and cost of manufacturing a wide variety of intricate designs and is beginning its usefulness in manufacturing as well as medicine.

## Summary of the Secular Effort to Extend Our Lives

These are but a few of the recent examples of medical research to demonstrate what the scientific mind can accomplish. These radical advances are progressing exponentially, and we can look forward to seeing the results of our efforts, if we will encourage the progress.

In order to effectively preserve and increase our knowledge base by allowing humans to live longer, we must face the facts of our physical design. Most of the medical research, as outlined above, that can potentially lead us into longer productive lives, involves our DNA, which, as explained previously, consists of a known molecular structure. Internally, signals are controlled by a stem cell's genes, located across the strands of DNA, and which carry coded instructions for all cellular structures and functions. The internal coded signals and instructions from DNA to all cellular structures and functions do not consist of matter, cannot be manipulated, or even completely understood. Perhaps they consist of energy, or perhaps there is a supernatural force involved which provides this intelligent code before we are born. Science admits that nature, or the environment, has no intelligence, therefore, any design or intelligence brought forth by nature, logically was implanted into the DNA of plants and animals by an external intelligent force. There is no other internal explanation.

Intelligence is order, and order cannot be realized randomly. This logic is the genesis of the Intelligent Design movement accepted by many scientists, who cannot bear to factor God into their belief system, for fear of losing their independence, and their perceived control over the world intelligentsia. Regardless of this paranoia that many of my colleagues possess, there is no logical reason that the "intelligent Designer" cannot be called "Creator," "Supreme Being," or "God."

One area of confusion in today's media and educational establishments is that they mistake *science* for *scientism, returning without reason to the philosophies of The Scientific Revolution, 1545-1727, rationalism, and empiricism, as described in Chapter 2.*

*Science* consists of the systems of knowledge of the physical or material world based on facts obtained through observation and experimentation. *Scientism* is the view that the hard sciences have the intellectual authority to give us knowledge of reality. (*Scientism and Secularism*, J.P. Moreland, Crossway, Wheaton, Illinois, 2018).

It is interesting that various disciplines of science are populated by different levels of belief in God. For instance, chemists are 41 percent believers in God, and physics/astronomy majors comprise just 29 percent. Between those extremes are biological and medical (32 percent) and geosciences (30 percent). It seems that the broader and closer to the living environment the science is, the more likely its scientists are to believe in God. Also, interestingly, the younger scientists (ages 18-34) are more likely (42 percent) than the older (28 percent) to believe in God. (*Scientists and Belief*, Pew Research Center, 2009).

Probably the most intelligent scientist of modern times, Albert Einstein, said in 1930:

*"The most beautiful thing we can experience is the mysterious. It is the fundamental emotion that stands at the cradle of true art and true science. He who does not know it and can no longer wonder, no longer feel amazement, is as good as dead, a snuffed out candle. It was the experience of mystery...that engendered religion. A knowledge of the existence of something we cannot penetrate, our perceptions of the profoundest reason and most radiant beauty, which only in their most primitive forms are accessible to our minds—it is this knowledge and this emotion*

*that constitute true religiosity; in this sense, and in this alone, I am a deeply religious man."* (Alice Calaprice, *The New Quotable Einstein*, Princeton University Press, 2005).

## Is There Natural Reality Beyond Our Present Knowledge?

### Technological Advances

If we force ourselves, and especially our government and our universities, to consider the option of committing to, and incentivizing, research which will help save our planet, as opposed to the more exciting, but futile, research into life in outer space, we could conceivably progress in the following areas:

- Returning the damaging portion of the ozone layer to earth, given that most greenhouse gases are heavier than air

- Insure weather control through disruption of natural or human influences on climate. If humans can cause a climate crisis, they can prevent it.

- Encourage energy research to develop more renewable and less polluting technologies in waste treatment, agriculture, clothing, shelter, transportation, recreation, and electronics.

One primary area of potential technological advancement is in the area of nanoparticles and how they can be harnessed to help us in our effort to save our planet.

### Quantum Mechanics

*Quantum mechanics* theorizes an *entanglement* between remote tiny particles. In other words, every particle in the universe could have an effect on other particles. These *effects* are too small to be observable. They are only discerned through *detectors*. But the theory that space emerges from networks of *entangled quantum particles*, and that entanglement forms the true fabric of the universe, indicates that space itself has disappeared, so the question of how communication occurs between distant particles, becomes easier to fathom.

The US Government and private industry is already spending millions to build commercial and research-grade quantum computers, which will solve complex problems much more quickly and accurately than conventional computers.

Einstein's theory of special and general relativity perfectly describes space, time and gravity in the largest scales of the universe, while *quantum mechanics* describes the tiniest scales, where there may be no space, time, and gravity. The whole may be more than the sum of its parts, if this theory is correct.

**Are these two seemingly opposite and contradictory theories brought together by God, helping to explain Creation?**

## Relativistic Quantum Mechanics – Neutrino Physics

The following explanation is taken from *The Theory and Experiment of Neutrino Oscillations,* by Michael N. Milam, Department of Physics, Princeton University, 2018:

*In the years following Einstein's landmark $E=mc^2$, a troubling discrepancy was noted in the context of nuclear beta decay. It was thought that beta decay was a two-body decay in which a neutron decays into a proton and emits an electron... But according to Einstein's theory, "the electron's kinetic energy should not vary and should equate to the difference in mass."*

In 1932, Wolfgang Pauli proposed that the solution of the missing energy was a *neutron*, later designated a *neutrino*. There arose controversy as to whether neutrinos have a mass at all. Recently (1985-1998), that mass has been determined to be positive, around 1/10,000 the mass of an electron, or 1/10,000,000 that of a hydrogen atom. Neutrinos are partnered with either electrons, or heavier versions of an electron, either a *muon* or a *tau*.

Neutrinos move at the velocity of light, and will completely penetrate and pass through any typical solid. Their source is not understood but is theorized to be the sun or even black holes.

## Fractons

The following is based on information furnished to the author from *The Dual Tensor Gauge Theory of Fractons,* a paper from the Princeton Department of Physics, by Michael N. Milam, Spring 2019.

*Fractons* are excitations of a quantum field that experience heavily restricted motion. *Fractons* exist in dipolar pairs and exhibit restricted motion, moving only in certain directions. Their fields apparently encode their behavior. To date, not much is known about *fractons,* and certainly not their purpose in the universe, but like Neutrinos and probably many other undiscovered particles, there is potential there for the advancement of civilization and the saving of our planet.

## What is the Potential for Understanding these Microparticles?

These are a few examples of our current advancement in the fields of *quantum physics,* which should demonstrate the potential for scientific and engineering advancement into areas heretofore unimaginable, but potentially could change and even save our civilization. Nature, for generations, has been explained and quantified by chemistry, physics, biology, microbiology, biochemistry, geology, math, and many other sciences. These sciences help us to understand the mass and energy of matter. *Neutrinos* are neutrally charged, therefore do not affect and are not affected by electromagnetic forces, and are therefore able to pass through great distances in matter without affecting or being affected.

Other tiny particles which have been discovered are quarks, bosons, muons, taus, gluons, gravitons, and of course photons, which include light and radio waves. Less is known about fractons, and there are undoubtedly many other microparticles still undiscovered; these are mentioned only as examples of what science is currently discovering, which may affect the future of our planet. Not much is known about these tiny particles, other than photons and electrons, especially their purpose and their source. As mentioned above, Einstein's Theory of Relativity relates energy, mass and the speed of light, and has been shown to explain physics in outer space and on earth, but seems to fail in the nano world. At the speed of light, at which these particles travel, and with their mass of near zero, their energy should also be near zero. But it is not, or they would never reach the speed of light. So, can we understand this newly discovered source of energy, or whatever we will call it, and be able to harness it to use for the benefit of civilization? Could we possibly use this new science to store and retrieve data, to replace our present generation of computers, which, as we have discussed, is rapidly using up our limited natural resources?

It is critical that we encourage the development of *quantum theory* and the science which develops from it if we are to survive on this planet.

Could they even, along with the entanglement characteristics of *quantum mechanics*, again be an indication of God's integral creation process of unification of all natural reality? I, being an environmental scientist, and not a theoretical or quantum physicist, cannot speak authoritatively on the subject of the use of energy from nanoparticles, but I can urge readers to be more involved with pushing these technologies, since the survival of our planet may depend on them, or something similar. God created this universe, and there is nothing wasted or without purpose. Even an atheist must admit that scientific facts explain natural truths, which can be harnessed to help civilization.

But in my opinion, if they could imagine and see, beyond their blinders, we would make much more progress. In effect, the atheists are today's emotional roadblocks, unwilling to accept, or even consider, a problem solution unless it conforms to their preconceived notions of reality. They are stymieing scientific advancement and should stand aside and allow those who are open to intellectual progress to have the freedom to lead us toward our planet's survival.

Science will never enable us to see God, but is it on the verge of detecting God's work of creating a universe which is completely integrated, with one supreme power having designed this integration, where all particles are interrelated and interdependent, as is nature? If so, is it possible that we are entering a fourth and final period of advancement of science, philosophy and theology, where, even though we can't discover God, we will be able to observe and appreciate His Creation through our effort at *detecting* the evidence of His involvement? After all, we certainly observe the evidence of God, every time we look closely at nature. Why can't "nano nature" provide us the same revelation?

**It would certainly be ironic if science, the perceived enemy of theology, were to be the mechanism for us to return to the ancient acceptance that the entire natural world is simply a demonstration of God's creation and the forces which explain it and hold it together.** If so, perhaps this fourth historical phase of the advancement of science, as it relates to thought, could be the impetus for a new and final revival, which will bring us face to face with ultimate reality.

## How Can Our Societal Actions Save the Air?

Of the three environmental realms, the air is the most sensitive to pollutants, since it provides the gas oxygen, that we, as well as animals, breathe, and the gas carbon dioxide, that plants use as their electron acceptor. As such, it is the realm that makes our planet unique in the universe for sustaining life. (*Earth's Atmosphere:*

*Composition, Climate and Weather,* Space.com, Tim Sharp, October 13, 2017). The air we have been given contains no nutrients or carbon-based organics that we can use as food. But the gases in the air are all critical for the protection, sustenance, and existence of the water and land realms.

The sun is our ultimate source of energy and the ultimate natural creator of our climate. Since the air is our insulation and our conduit for that energy to be transmitted to our climate for control, there is a harmony in the balance of gases in the air that can be disrupted by civilization.

The air is a resource for living organisms, since its elemental mixture is naturally perfect for organic existence, but is not appropriate as a receptor for the additional residuals of the unnatural pollutants that we as a civilization emit from activities in the air, on the water or on the land.

If inorganics or organics are discharged into the air, the air acts as a distributor of those pollutants and not a treatment mechanism. Because of the tendency to widely disperse pollutants to distant areas, our atmosphere must be controlled through unified practices.

**Therefore the protection of air is of worldwide concern. There must, obviously, be coordination and cooperation between all nations in order to develop implementable practices that truly preserve the natural state of the air.**

Chapter 9 describes that natural state and the natural percentage of each constituent. The balance of our air is so fragile that in order to sustain life, the natural composition of air must be maintained. One of the more sensitive constituents of air is carbon, therefore, the carbon balance is essential for life. Venus has about 100 times as much carbon as the earth and suffers from a runaway greenhouse effect that has produced an average surface temperature of about 891°F, compared to about 81°F for the Earth. Mars, on the other hand, has about 1/100 the carbon of the earth (*Hyperphysics*,

Georgia State University, 2016), with corresponding cold temperatures (average temperature -81F).

As consumers of energy, we directly add $CO_2$, methane, oxides of nitrogen, and sulfur oxides to the atmosphere when we burn materials in our industrial activities, and in our automobiles, heaters, stoves, ovens, and water heaters. We, likewise, contribute to the addition of these materials to the atmospheric carbon cycle when we breathe, use electricity, buy products, which are manufactured and transported, or use public transportation.

## Carbon Dioxide Control

Industrialization of our society has caused a continuing increase in the concentration of $CO_2$ in the atmosphere (see *Preface, Does Our Planet Need Saving?*). The EPA has listed $CO_2$ as about 82 percent of all emitted Green House Gases (ghgs), which can contribute to the greenhouse effect by limiting back radiation of heat coming from the earth. Many other ghgs have been eliminated or reduced by regulations, but $CO_2$ is the natural emission from animal breathing and the product of combustion and, as such, is much more difficult to control. There are four general methods of reducing or eliminating the $CO_2$ increase in the atmosphere:

- Minimize combustion by substituting alternative energy production sources such as nuclear, hydrogen, wind, solar, geothermal, and hydraulic (waves and stream flow).

- Use renewable carbon sources such as methane, wood, and vegetation (biomass) instead of non-renewable sources such as coal, gas, and oil. Even though the amount of $CO_2$ added to the atmosphere is roughly the same, renewable sources not combusted will eventually emit $CO_2$ upon decomposition if not harvested. Where non-renewable sources add net $CO_2$ to the atmosphere for the first time when they are burned, rather than recycling $CO_2$ that is already in the carbon cycle. Therefore, there is a net $CO_2$ increase in the atmosphere only from non-renewable sources.

- Increase the efficiency of combustion so that less $CO_2$ is emitted relative to organics destroyed.

- Capture and secure by sequestering the $CO_2$ emissions in water, soil, or vegetation. The $CO_2$ is captured and stored (not destroyed or used) in water and soil (even deep injection) and beneficially used only by vegetation. These sequestering reservoirs are called sinks.

According to *Carbon Sequestering* ( An on-line course by Lee Layton, based on the U.S. Department of Energy report *Carbon Sequestration Research and Development,* December 1999), the following sources and sinks are annual estimates of the global $CO_2$ cycle during the 1990s:

| $CO_2$ Source | GtC* | $CO_2$ Sink | GtC* |
|---|---|---|---|
| Human and animal respiration | 60.0 | Vegetation (photosynthesis) | 61.7 |
| Deforestation | 1.4 | None | 0 |
| Fossil fuels | 6.0 | None | 0 |
| Decay of ocean vegetation | 90.0 | Ocean uptake | 92.2 |
| Source total | 157.4 | Sink total | 153.9 |

* Billion metric tons of carbon from $CO_2$ = gigatons carbon (GtC)

These estimates indicate that an estimated 3.5 GtC per year enters and remains in the atmosphere.

Recent research attempts have been made to quantify the loss and gain of $CO_2$ in the ocean due to temperature variation, but these estimates are difficult to substantiate because of the variability of ocean temperatures due to geography, seasons and currents.

The ocean is estimated to contain about 40,000 GtC of $CO_2$.

If vegetative sequestration remains the same, the oceans would only increase $CO_2$ by 8.75 percent in 1,000 years, but because of the logarithmic nature of pH, there could be a gradual lowering of ocean pH caused by human $CO^2$ emitting activities (deforestation and fossil fuels), which could affect the growth of coral reefs and consequently, a loss of local ocean biomass.

The vegetation sink for $CO_2$ has a potential through management policies to affect the sequestration of human-caused $CO_2$ emissions. Because of the increase in vegetative growth in the United States in recent years, there is a trend of a net increase of $CO_2$ sequestered; but in the entire world in 2000, there has been a net annual reduction of $CO_2$ sequestration twenty times larger than the U.S. increase (*Trend Estimates of Land-Use Sequestration*, U.S. EPA, 2019). So, if vegetation is less available as a sink, except for a small amount entering the soil, the emitted $CO^2$ can only be sequestered in the water and the air.

EPA has estimated that through changes in agricultural soil and forest management, tree planting, and biofuel substitution, the U.S. could increase its vegetative sequestration by 30 to 90 percent. In 2002, about 12 percent of the total $CO_2$ emitted in the U.S was sequestered.

It should be emphasized that young forests use more $CO_2$ per acre than old forests. Young trees, like teenagers, metabolize more rapidly, using more $CO_2$, and old forests have more products of decay which produce $CO_2$. It has been estimated that a tree takes up about 2.52 pounds of $CO_2$ per day (*Tufts Climate Initiative, "Sequestration: How much $CO_2$ does a tree take up?"*) This means that roughly one twenty-five-year-old tree is required to take up the $CO_2$ exhaled by an average man, which is estimated at 2.3-2.5 pounds per day.

Quite amazingly, a new United Nations scientific report, states that *But if people change the way they eat, grow food and manage forests, it could reduce the chances of a far warmer future.* Perhaps, even the UN realizes that their traditional emotional arguments are beyond acceptance.

## Other Greenhouse Gases

In recent years, fluorinated hydrocarbons in the form of refrigerants, have become an air pollution concern.

Due to wind and thermal activity, dispersion of these gases can be regional or international in scope. The dispersed pollutants can remain in the air permanently or at least for long periods of time, especially if they are vapors or gases. If the pollutants are lighter than air, or close enough to the specific gravity of air, they can rise or be carried higher in the atmosphere by wind and thermal currents. If the pollutants are particulates, or heavier than air, they can settle, or be carried by air movement, back to the surface. Therefore air pollution may be dispersed in the air or carried to bodies of water, or the ground. For example, the 1883 eruption of the Krakatoa Volcano raised the average global temperature by as much as 2.2 $^0F$ until 1888 (Wikipedia).

**How can our Societal Actions Save the Water?**

The water of our planet is the next most sensitive realm and, although not as much as the air must still be protected from pollution in order to preserve its use for navigation, recreation, irrigation, water supply, and fish and wildlife. Water quality is a regional rather than a national or global concern. Historically in the United States, and still, in some countries, water bodies are even used as sewers to transfer pollution away from civilization, but still in the region. The requirement for the purity standard of water varies with its use, but the most rigid requirements are for water supply and fish and wildlife. Without sufficient water quality to sustain these uses, civilization and nature would potentially cease.

If there were no living organics on the earth, and the water cycle remained as is, the water bodies would collect and distribute the inorganic dust and particles from the land. The presence of living organics on the land, and their conversion to non-living organics through the natural life-death cycle are also collected and distributed by natural water drainage systems.

Certain living organics accumulate in the water bodies and provide nutrients and food for aquatic life and flying wildlife. Therefore, water is a sink for inorganics and organics carried to it by the water cycle. The inorganics provide no benefit to the water as a food source except as nutrients. The waters of the earth are the largest receptor and container of organic life on the planet.

The water provides very little treatment of waste organics flowing into it, other than allowing bacterial and aquatic life removal of pollutants through the presence of dissolved oxygen in the water.

## How can our Societal Actions Save the Land?

The least sensitive of the three realms of our environment is the land. Land pollution is usually only of local concern. The land is the source of most raw materials used by industry for manufacturing of consumer goods. Once a constituent of the land is removed, it is lost forever unless it is replaced in its original position.

The land realm is the source for most of the earth's food supply for land-based living organisms. This food supply is both naturally occurring and human developed.

**The current practice of storing waste products in landfills provides little or no treatment of these products and is limited by the amount of land available for this purpose in the future. Waste products entering a landfill are typically stored in cells, not mixed with natural soil bacteria.**

The other critical issue concerning land is the permanent removal of limited resources from the land, which are never replaced. These resources below the surface of the ground are both organic and inorganic. Once these resources are removed, it is virtually impossible to replace them at the source (see Chapter 11). Organics in the soil have their source from vegetation and include peat, oil, gas, coal, and diamonds, the result of pressurizing organic matter over time. Except for diamonds, these resources should be considered as not only non-replaceable, but as adding carbon to the atmosphere when converted into energy by burning.

There are three types of land resources on the surface of the land; vegetation, soil and surface minerals. These resources, especially vegetation, are generally replaceable and do not add to the carbon cycle over time.

The land is the original source of all of the organic material on the earth except for a few organic gases found in the air. Much of the inorganic materials found in the air and especially the water, have the land as their source.

## Is Diversity a Key to Saving Our Planet?

Many elite educational establishments, political parties, private clubs, and churches advertise diversity as their primary goal. Does this goal work toward saving our planet? *Diversity* is defined as *dis-similarity, difference, variety* (Webster's). Both Creation and evolution teach *diversity* as a logical result of animal procreation with time. When we pride ourselves on being diverse, we must ask ourselves: diverse from what? Is it ethnicity, religion, geography, prior education, intelligence, mental stability, or what? It seems that schools pick which diverse students they can handle, and reject those they can't. That is false diversity.

Biblically, diversity is when everyone does his part in order, rather than chaos, as when everyone does what is right in their own eyes. Our goal today in society, education, and employment seem to be to defeat natural *diversity,* and force all humans to be "peas in a pod." Many private and religious schools are accused of that, and rightfully so. Likewise, especially today, many of these schools use their success at diversity to brag about their social sensitivity and forward thinking.

For years, public schools, being required by law to provide an education to all citizens, and some private and religious organizations, have led the government to support and regulate diversity, to the point that we basically all agree with the concept. But their key to success is to provide special education to certain diverse students.

Most of us have experienced an awkward situation where we seem trapped in a mini-culture of specialists and having no clue as to what they are discussingt. That is the effect of "main steaming" neodiverse and sometimes economic or racially diverse students into the rigid and necessarily limited curriculum, and requiring them to produce educational success. It ends up being counter-productive, cruel, and ruins lives which otherwise could be joyful and fulfilling.

As humans, we are naturally competitive. We have athletic competition separately for different age groups, and the Special Olympics for the physically and mentally handicapped, and it works wonderfully. Have you seen the joy in the face of a Special Olympics competitor? It would be absent if that individual were required to compete in AAU Track.

So academics, being much more critical than sports, must catch up and dedicate more funding, more teachers, more space, to educate **all** humans at their age and capability level. Not at just one level, but with multiple levels and with teachers qualified for varying impairments and challenges. We must "leave no one behind", but not necessarily blend everyone into a fixed curriculum. Justice is to raise up the vulnerable, and their "raised" condition may be different from the norm, but not lesser. Their norm is typically as important to society as other more accepted norms. The exception, of course, is mental or physical impairment which could prohibit a person from contributing to society. They are the truly vulnerables, and the ones society must protect and encourage. Then we will usually find that these souls have more love to offer than we have ever experienced, and realize that love is much more important to society than education.

**We should celebrate diversity, not force it. It will come naturally if we can define from what we are diverse.** The Nazis defined that base level as the perfect Arian, and murdered many of those not meeting that criteria. So our challenge for education is to define a range of academic ability; and those candidates who are diverse from that goal, should be educated at a different level to become functioning members of society. Sports and academic grading differentiate between ability, as does free enterprise. **But society includes positions for a wide variety of academic abilities and our job is to match those opportunities with our education and employment systems.** Also, we must ask why we rate academic ability higher than physical or mechanical ability, or vice versa. **Some professions require gifts other than academic proficiency, gifts that society could not do without. Those gifts should be respected as diverse gifts, not diverse people, not lower than academic gifts.**

One of the strongest opponents of the return of education to rational thought and critical thinking, is the educational establishment itself. More and more, especially in the "elite" high schools and colleges of our country, the goal of the school is not education, but diversity. It seems that if the purpose of education is to produce graduates with more knowledge and wisdom, the goal would be to accept only qualified applicants, and have a corresponding system through which academically under achieving applicants have an opportunity to grow to an equal quality in their specialties, through their own effort, and in their own timing. **If one purpose of education were to produce graduates qualified to perform as productive members of society with gifts other than stereotypical knowledge and wisdom, the goal would be diverse education, not a** cloning of all diverse students into one preconceived view of acceptance.

The current system of reducing educational effectiveness to a common denominator will end up destroying the entire system. After all, most every student is intelligent enough to realize that if he or she can graduate and get a job with no effort, why stress themselves; and if they are forced to take courses in subjects of no interest to them, why even try? **If our country wants everyone to perform and be compensated exactly the same, there is no incentive for excellence, and no motivation to discover and accumulate facts that will save the environment and our planet**. Also, our country will end up with no manual labor or service pool, and cannot function. **So the conclusion is that we must return our country to a merit based diverse educational system, containing opportunities for all positions in society, open to all; give the initially "unqualified", the opportunity to be trained and educated, or better, determine how they can best fit in; what they are qualified for. In other words: our goal must be diverse education, not just diverse students. More people in society are needed for the so called "menial" jobs, which are not menial at all, but completely necessary for our society to function. Otherwise, our educational system, along with the possibility of developing new methods and products that can allow us the save our planet, is an unrealizable dream.**

## Practical Examples for Society Improving our Environmental Stewardship

The following are several examples of how we as part of our society can aid directly or indirectly in our responsibility as stewards of the environment:

1. Support efforts such as the Comprehensive Everglades Restoration Plan, a program managed to restore natural stormwater distribution.

2. Support privately granted Wilderness Preserves which deed tracts of land to a permanent trust which prohibits their use for anything other than the preservation and enjoyment of the natural state of the land.

3. Promote wildlife and forestry management to enable people to beneficially use their property without depleting its resources.

4. The management of invasive species of animals and vegetation can be locally accomplished or regionally accomplished by governments or organizations. Examples are the control of Kudzu, the beautiful but destructive lion fish, the python and tree blights.

5. The consideration to allow the less fortunate or less trained equal opportunity to enjoy the aesthetic pleasures of the environment, whether on private or public property, while balancing access with survival of the environment. Too much access to government property is denied to the citizens whom the government serves, in the name of protection. Why not just require training for access, as in the granting of a Driver's License?

6. Unselfish sharing of our property, neighborhoods and towns for relatively undesirable environmental related projects such as tree farms, windmills, solar farms, landfills, sewage treatment plants, etc. The old NIMBY (not in my backyard) argument does not seem to be a reasonable concern in a community. We have spaces in our houses such as bathrooms, trash areas and laundry rooms, which are not exactly pleasant areas to hang out, but we can close them off. Can't we do the same thing with our community land instead of whining NIMBY?

7. If we can potentially resolve our educational crisis through motivation and incentivation, we need to develop a system

of retrieval and storage of the scientific knowledge we now, or have in the past, possessed, and make it available to research entities across the world. With this basis of facts potentially available, we can possibly reach the state where we are able to store all these facts of the environmental reality in sub microscopic, or sub atomic space. We could then develop a system of retrieval, manipulation and implementation for these facts to be of beneficial use; a system of artificial intelligence comparable to our brains, but controlled by our minds. Each fact stored must be locatable as it relates to an input goal, and be manipulatable in conjunction with other retrievable facts in order to arrive at the most efficient, economical, just and fair conclusion, for advancement to be realized.

This potential can therefore analyze the understanding of our minds and our environment of air, water and land, and determine how we can work together for maximum sustainability.

The weakness in this potential is that many of these stored facts are living organisms, not currently subject to control of humans, or facts subject only to extra planetary forces, such as the Sun. These organisms and outside forces were either created as such by God, or evolved, against all scientific laws, from inorganics, or extra planetary forces, from nothing. So one limitation of our model to save our planet is that "nothing" is not a fact. Therefore, we can't store it, retrieve it or manipulate it. For instance, we cannot store the idea that God does not exist, since it is a negative.

Unfortunately, in a way, that limitation applies to all living organics, since they are programmed by their DNA. Undoubtedly we will discover much more about DNA in the future, but the limitation remains at the present time; that a living (there is scientific debate about whether DNA is a living organism) organism, such as DNA, cannot, and never will be able to be constructed from "nothing". Artificial intelligence cannot explain or construct something out of nothing, so the conclusion of impossibilities will be a waste of valuable potential in the future.

**Science, and even theoretical physics is destroyed, or at least limited, if something can be created from nothing. The only practical and reasonable answer to that conundrum, is that there is a supernatural, or spiritual reality which is beyond science, pre and post Creation, which is in ultimate control of our natural world. And if so, we can never completely understand that unnatural reality. It is beyond our understanding, therefore out of our control, and the only way to co-exist with it is to submit to it.** This is what absolutely drives atheistic scientists (an oxymoron)mad; to state that anything is beyond their understanding and control. When your entire life is devoted to the understanding of all natural things, to be able to admit that there is truth beyond, is very humiliating, and requires the removal of opaque blinders and their very limited visual range, to accept the supernatural. A scientist is not a true scientist if he (she) is so arrogant as to truly believe that they are beyond scientific law and reason (conservation of mass and entropy, for instance). See *The Future of Species* by the author for further proof

## Summary of how Society can Save our Planet

In summary, civilization must be responsible for two areas of concern if we are to improve our environmental stewardship: 1) **To conserve non-renewable resources by replacement in the land, reusing or recycling all inert materials extracted from the land which aren't burned, treating all flammable organics in the ground as non-renewable, and developing renewable energy sources.** In effect, treating every inorganic or organic removed from the land as lost to society, unless and until it is replaced. This can be accomplished by minimizing waste and substituting the use of non-renewable resources with renewable resources; and 2) **Establish a worldwide goal of returning the air, water and land to the chemical composition it would possess if there were no humans on earth.**

**A rational solution for the entire environment is to prohibit all discharges into the air which do not meet ambient, unpolluted air quality conditions. This is the theory behind water and land pollution control, why is it not used for our most sensitive environmental realm?**

If this change is implemented for all permitted discharges into the air, water and land, all polluters, from individuals to the largest power plants, will be forced to re-use, recycle and/or minimize waste.

**What if, instead of a "carbon tax", every discharger of $CO_2$ were required to plant an equivalent number of trees locally; the $CO_2$ uptake of which would equal their $CO_2$ discharge?** It has been estimated that a tree uses about 2.52 pounds of $CO_2$ per day, and that young trees use more than mature trees. (*Sequestration: How much $CO_2$ does a Tree Take Up?* Tufts Climate Initiative). **That figure will allow us to determine the number of trees which must be planted per year in order to sequester the $CO_2$ from each emitter.**

**What if every manufacturer were required to accept returns of the waste from the products of their manufacture?** That would force reuse and recycle from the manufacturers of clothing, buildings, recreation equipment, transportation equipment and electronics. Data varies across the country, but In my city, Nashville Tennessee, as an example, in 2016, 72 percent of commercial waste was landfilled, 23 percent recycled and 5 percent composted. Of the construction/demolition waste, 99 percent was landfilled and 1.5 percent was recycled. Residential waste was landfilled as 87 percent of the total, recycled 5 percent, and composted 7 percent. (Metro Nashville Public works Master Plan, CDM Smith, 2016).

**What if every home, restaurant, school, etc. were required to compost, and reuse all non-composted waste food in animal food products or directly as animal food?**

**What if more non-renewable, biodegradable or non-biodegradable resources extracted from the earth were taxed more, and the income used to develop replacement technologies?**

**In all of these suggestions, manufacturers will complain and consumers will have increased prices, but what is the environment and the future of our planet worth? And who should pay this cost if not those who use and manufacture the products derived from the planet and not returned? This seems to be a much more rational approach to saving our planet than the current emotional pressure that is limited to $CO_2$ control.**

When we objectively analyze our potential to save our planet, several glaring deficiencies arise:

- The public is educated in complaining about our water and land quality, but not in preserving it.

- EPA sets most regulatory standards based on affordable performance, rather than risk based standards. This should not continue if we are to survive environmentally as well as economically.

- Reuse and recycle requirements should be intensified with tax incentives, etc. to encourage compliance.

- Regulations should require that manufacturers require return of all used products for recycle.

- The EPA standards must take into account the coordination between industrial discharges and receiving stream assimilative capacity; something they have attempted to do, but need radical improvement. An industry should be forced to move or meet the standards for its receiving stream. The environment should not be punished in one locality because it is more sensitive in order to standardize regulations across a specific industry.

- The EPA standards must be scientifically set, based on risk to the environment and human health, regardless of when the industry was established or where it is located.

- Industries should plan for future regulatory changes and update their production and pretreatment procedures.

- Each individual home must be considered a source of air, water and soil pollution, and regulated by local enforcement agencies under federal guidelines.

- Pollution sources must be given adequate time to plan and implement new requirements.

- There must be limitations placed on lawsuits against enforcement agencies, which only serve to delay compliance.

- The EPA should be above politics and transferred from the Executive Branch of the government to the Legislative Branch, or preferably into a non-governmental, publicly owned organization, such as a Union of States.

- Until the EPA can become independent of politics, more environmental control and power should be given to the states than the Federal Government, but under uniform federal standards.

- Standards for air pollution control should require that no vapors are to be returned to the air at a lower quality, or chemical constituency than existed when they were removed.

- Standards for water pollution control should require that no wastewater be returned to the water at a lower quality, or chemical constituency than existed when it was removed. Also, standards should require that wastewater be returned to the stream from which it was removed. This would encourage, or force re-use of treated water. This is basically the "Non Degradation" policy that EPA was directed in the 1970s to enforce.

- The EPA RCRA Standards applicable to landfills and land disposal of solid wastes should be completely re-written to force landfills to separate and map storage areas, and attempt to return specific wastes to their original source.

- Chapter 10 gives recommendations, which can be used for societal or individual residuals accounting.

*Chapter 14*

# How We as Individuals Can Save Our Planet

## Introduction

A suggestion for each of us, as individuals or as part of industry, commerce, or government, is to develop a comprehensive, cradle-to-grave method for examining the extent of the residuals that we, in our position, can control, in terms of pounds per year. Then we can determine if, and how, these residuals re-enter the environment. The following are recommendations for any level of responsibility, in organizing a rational formal program to reduce environmental pollution.

## Residuals Accounting

The State of California has, for decades, followed an environmental policy to require industries to examine their total waste discharge and to commit to its annual reduction. This policy requires the industry to justify that the raw materials it purchases cause the minimal environmental pollution, and whether they can produce less polluting products. They must also analyze their manufacturing process for minimizing pollution, and even to see if the final disposal of their products by consumers can produce less waste.

This philosophy, without the complicated paperwork with which the government is obsessed, can be used for us, as individuals, in our effort to be environmentally responsible. I would recommend the study and possible adoption of a process similar to one developed for greenhouse gas emissions called *The GHG Protocol for Project Accounting,* which offers an objective approach to accounting for residual discharges of all types. This Protocol was published by The World Resources Institute of the United States and The World Business Council for Sustainable Development, of Geneva, Switzerland. This group is a coalition of around 200 industries, governments, and non-government organizations launched in 1998 to develop internationally accepted greenhouse gas (ghg) accounting and reporting standards.

The following is a recommendation for using these principles for anyone desiring to be an environmental steward:

1. Write a *Missions Statement* expressing in a sentence or paragraph form, the *purpose of your residuals reduction effort.*

2. Determine the boundaries of your plan for direct and indirect uses.

3. Estimate the pounds per year of waste discharged under your control for direct and indirect uses, and the current discharge amount for comparison.

4. Determine discharge lowering goals.

5. Document the intended method of residuals reduction, the amount per year of reduction, and the ultimate goal.

## The Mission Statement

The Mission Statement should be a simple expression of what you are going to do, agreed to by all involved parties. A possible statement could be: *The goal of this plan is to follow rational scientific principles in assuring that we lower the total amount of residuals under our control, which are returned to the environment.* Since this statement is the first step of the accounting process, the amount and timing of the reductions has not been set.

## Plan Boundaries

The boundaries of your plan are two fold – the geographical boundary and the organizational boundary.

## Geographical Boundary

The geographical boundary is the physical area from which discharges to the environment occur. This, from a family or individual standpoint, can be your property, or a portion of your property, whether owned, rented, or leased.

## Organizational Boundary

The second, and overlapping boundary, which should be considered, is the organizational or indirect boundary. This boundary includes the environmental discharges from products and systems which you own or control. Controlled products or systems could be rented, leased, or sub-contracted. They could include transportation off-site for errands, deliveries, etc. Discharges from these products or systems can be included, or excluded, from accounting, but should be clearly understood when comparing results.

## Life Cycle Analysis

The indirect and/or the organizational boundary could literally be the world, if you complete your accounting on a life cycle basis. The philosophy of a life cycle analysis is to include all of the indirect discharges associated with the product or system, such as those associated with manufacturing the raw materials, shipping the raw materials, fuel and energy use during manufacturing, fuel and energy use during the life of the product, and the ultimate disposal of the product.

Most practitioners of residuals accounting do not use a life cycle analysis, since theoretically, each supplier, transporter, user, etc. is responsible for his own accounting, and if everyone practiced life cycle analyses, there would be a huge overlapping of reporting.

In our personal, as well as in our corporate responsibilities, the farther groceries, supplies, and other household needs are produced from their point of sale, and the farther we travel to shop for them, the more air emissions are required for delivery. The more dependent we are on outside supplies and services, the more our "cradle to grave" responsibility increases. So we should never completely ignore life cycle effects on the environment.

One example of that life cycle approach to environmental protection is the current "feel good" trend of using organic foods, cremation, and electric vehicles. A life cycle analysis would require the consideration of the environmental effects of manufacturing, shipping, energy, and disposal. Our government, in the case of electric vehicles, is subsidizing the cost, and requiring taxpayers to pay for a possibly more damaging solution to saving our planet. So, this cost of subsidizing should be considered as an indirect cost, since it is paid for by other citizens. See Chapter 10 for more information on this subject.

## Accounting for Pollution From Our Everyday Activities

As a civilization, we use burning to provide heat for transportation, comfort, and cooking, and also for the destruction of unwanted trash and garbage. The potential pollutants from burning include carbon dioxide from the incineration of hydrocarbons such as wood, paper, food, and cigarettes.

We, as individuals, may influence the total environmental pollution from our residences through our employment, outside activities, consumerism, and recreation. This adds up to a potential of 100 percent.

## Reuse and Recycle

If we consider that all residuals of society are initially derived from our air, water, or earth, and after use, must be returned to the air, the water or the earth, virtually all of our air pollution has a source other than the air. An imbalance is therefore forced on that part of the environment. The only solution to this imbalance is to return all residuals to the point of their original source. Because of our global economy, that goal is impractical and unachievable. But it is potentially achievable to consider the reuse or recycle of all residuals. This can be accomplished, again, by individuals at their home or place of employment. "Give away, rather than throw away," "Your trash is someone's treasure" are positive methods of recycling.

One possibility of encouraging reuse and recycle is for all industrial sources to be required to accept the used products they manufactured, back as raw materials.

## A Summary of Practical Ways to Save Our Environment as Individuals

We have discussed our Culture Wars, and the need for transferring our energy and finances to the more critical issue of saving our planet. Our minds, their limitations and their potential in saving our environment must be harnessed in order to accomplish this objective. Hopefully, we are encouraged that we can indeed accomplish that goal. We have also examined the environment itself in detail, in order to discern its strengths and weaknesses, which affect the survival of our planet. Hopefully, these thoughts have motivated us to do our personal part to preserve the air, water, and land for future use, while minimizing the use of our limited natural resources. But under the current practices and technology, that will not be enough to save our planet. We must develop new technologies considering these limitations, convince our regulatory bodies to allow experimentation, and our government into funding and sharing new and radical research and development, in order to survive.

Why not really think outside of the box, put our emotions on hold, and consider using only truly renewable resources for our energy needs? Practical? Perhaps not, but certainly possible. Today we depend almost totally on non-renewable energy such as non-hydro produced electricity, natural gas, and petroleum. What about natural renewable sources of energy? What about excess unused sources of energy? Natural renewable sources include solar, wind, water movement, geothermal energy, and vegetation. Each has its limitations. Solar, only when the sun is shining; wind only in certain areas at certain times; water movement limited to the ocean, streams, rivers and underground channels; geothermal energy according to depth into the ground; and vegetation according to its management. Each of these energy recovery systems also requires manufactured equipment to implement, but are all underused.

**Every building's water supply and sewage discharge includes excess energy in the form of liquid movement, which can be harvested.**

*The amount of heat within 33,000 feet of Earth's surface contains 50,000 times more energy than all the oil and natural gas resources in the world* (*How Geothermal Energy Works*, Union of Concerned Scientists, 12/22/14).

Renewable vegetative energy requires management, harvesting, land use, and air pollution control. But all non-renewable energy resources have larger disadvantages and are typically more dependent on transportation, using even more energy. We should further explore the giving of tax credits for tree planting and renewable energy systems.

*Chapter 15*

# What is the Ultimate Fate of Our Planet?

## Introduction

No one knows what will ultimately happen to our planet unless they accept that the words of the Bible are true when they prophesize that the universe will be replaced by Heaven. Without that assurance, there seems to be only two other options, unless we as a civilization are able to save our planet by restricting pollution and the use of all of our available natural resources: we revert into an uninhabitable planet, or we return to nothing.

Natural reality can roughly be divided into three spheres. The largest being space outside of the earth; the smallest being the "world" of microscopic and sub-microscopic matter and energy outside of our body, and in between, our observable planet, including our bodies.

Interestingly, it seems that each of those spheres contains a similar number of objects.

That is the extent of our current knowledge. What does the future of science hold in the discovery of more natural reality, and more about the reality we know? How many zeros" will we add as we discover far outer space, smaller and smaller sub microscopic particles, and more about nature, including our bodies?

Since the fall of man, as described in the Bible, or to atheists, the beginning of humanity, humans and nature have been enemies, bent on each other's destruction.

But even as natural enemies, humanity is completely dependent on nature, the environment, for its survival.

In Chapter 1, we posited the likelihood of the existence of two parallel realities; the natural and the spiritual. If that duality of existence is true, we must admit that the following conclusions are true for the natural and spiritual realities as they relate to saving our planet:

## The Natural Reality

The natural reality has been created with inorganic mass and energy, which can last forever, and organic matter which may, or may not, last forever.

The natural reality has been created with living organisms which have a limited life span.

In addition to natural matter which has mass, the earth was created with energy, which is theoretically interchangeable with mass, and has a finite total quantity which cannot be increased or decreased. This mass and energy can be changed, but it can neither be created nor destroyed (The Law of Conservation of Matter). In other words, there is a mass balance of mass and energy in the universe.

The Second Law of Thermodynamics states that the *entropy* of an isolated system never decreases over time. Non isolated systems may lose entropy if their environment has an entropy increase to give a balance in the form of quantities such as volume, pressure and temperature in aggregate. In non-isolated systems, in a reversible chance driven system, like evolution, entropy does not change. Irreversible systems, like creation, always increase total entropy. Entropy is an expression of the randomness or disorder of a system, as opposed to order, as expressed in mathematics or the laws of science. If evolution were true, therefore, our universe would be one of order with a balance of evolved and devolved matter. But Darwin, with his poor understanding of science, invented a system of evolution always improving, while we can observe "unevolved" examples all around us. And yet, we have never found evidence of a human devolving into an ape. We do see evidence of entropy every time we look at our fingerprints or our face, or when we throw individual letters into the air and expect that they will land in some sort of order.

These natural realities can be proven scientifically.

## Spiritual Reality

Likewise, the following conclusions must be true for spiritual reality:

- The spiritual reality preceded and created the natural reality for its own purposes, or according to the Law of Conservation of Matter, the natural reality would not exist.

- The spiritual reality has the ability to control the present and future of the natural reality.

- In a spiritual reality, without a universe, there is no way of measuring time, and therefore, no time.

- Since the spiritual reality created the natural reality, from the natural perspective, the spiritual reality is omniscient (all-knowing), omnipotent (is all-powerful), and is omnipresent (is always present everywhere).

- The spiritual, and therefore the natural realities, must be under the control of a force or being, historically called *God*.

- These spiritual realities cannot be proven scientifically, but were written to us in the Bible, transcribed by humans influenced directly by God.

The purpose of the rest of this chapter is to step back and examine each of the two realities in order to determine whether the capabilities of our minds and our environment afford us the possibility of saving our planet through environmental stewardship.

**Our responsibility is, therefore, to determine how long the current rate of damage to our planet can last, and still support human life, and whether we can extend that projected life of our planet.**

Here are some thoughts and questions we must consider, and answer, if we are to save our planet. After all, our survival must be related in ways we do not yet understand, to these larger and smaller spheres of interest, as well as to observable nature.

## Outer Space

There is much pressure on the U.S. Government from NASA and much of the scientific community to explore outer space, especially to attempt to find life. We spend billions a year listening, looking, visiting, and theorizing about a seemingly sterile outer space, in the hope of finding signs of life or, better yet, an advanced civilization. Will that effort and that expenditure help us to save our planet? Can something in the natural realm outside of our planet help us to live longer or affect our future survival? The truth is that we don't know, but are we willing to allocate public funds toward that goal, rather than to the discovery within our planet's nature, of solutions to saving our own world and our own future? There is no question that we need a presence in outer space to defend ourselves; for communication, observation, and control of worldwide events; and to potentially extract valuable inorganic minerals, but beyond that, our exploration may be wasted effort.

There are many questions about outer space that have not been answered, which may not affect our future, but certainly are of interest. Perhaps these questions should be addressed by joint non-government and government organizations, and not divert only our tax dollars from more important research that can more reasonably save our planet.

## Observable Nature

As discussed in Chapter 13, in medical science, there has recently been a revolution of advancement in the use of living organisms, as opposed to inorganic chemicals, for the repair, replacement, and healing of human parts and organs. This dramatic advancement appears to objectively indicate, from a secular standpoint, as is accepted from a Biblical standpoint, that the purpose of the planet, and indeed the universe, is to provide habitation to humans. If relatively sterile outer space does not provide the environment and the natural resources required for our sustainability, they must be provided within our planet. Since the goal of survival is human survival, medical science must hold the key to the saving of the purpose of our planet. So, to prevent our planet from becoming uninhabitable, or returning to nothing, we must do our part, and force those in authority to do their part to continue to make medical advancements using living organisms.

## Nanospace

In science, there are sub-microscopic nano particles in the universe which are detectable, but not observable. Most, and probably all, have mass, and all have energy. We do not understand the purpose of most of these particles, but they could have tremendous potential in the understanding of our planet and its survival. As discussed in Chapter 13, the harnessing of the energy of these particles could potentially solve our natural resources limitation, but it is unlikely to solve our pollution issue.

## If Not Outer Space, Nature and Nanospace, What is the Answer to Our Fate?

If our goal really is to save our planet, rather than to win the Culture War, or to discover something interesting that we could name after ourselves, we must determine what parts of the environment really affect our longevity and the longevity of the rest of the living organisms on the earth. All other effort may be challenging, and may be interesting, but may do little or nothing to save our planet.

So, my conclusion is that yes, the scientific mind can save our planet *until* God wills otherwise. **God created humans to do His will. We don't always know His will, but He has given us in this generation, enough knowledge, enough ability, and enough resources to potentially save His Creation, for as long as He allows. Now our job is to use that knowledge and ability, and those resources to make a difference.**

The first step toward that goal is to understand the environment that God gave us, as described in Chapter 9, and to understand the minds that God gave us, as described in Chapters 3 and 6. Why would He create a complex and amazing environment, including us, if He knew we were going to destroy it? **An unlimited God, to whom time means nothing, knows exactly what will happen in the future on this planet.**

Assuming that God has given us this opportunity in this generation, the next step is to combine our resource, the *mind*, with our challenge, the *environment*.

## If We Do Nothing, What Then?

**One of the wonders of life is that it is a renewable resource.** What does it take to produce and sustain life? Two animals or plants of different sexes, food, as an energy source or electron donor, and either $O_2$, nitrate, nitrite, sulfate, or $CO_2$, as the electron acceptor. This is radically different from the non-living inorganic portion of our planet, which is limited by definition.

**So once we use up our inorganic or organic natural resources, we cease to exist.**

As scientists will tell you, the environment of our planet is very sensitive. If we raise the percent of the $O_2$, in our atmosphere by even 2.5 percent, we as a human race will soon cease to exist. Likewise, if we lower the $O_2$ percent by even 1.5 percent, we will soon cease to exist. So one prediction of the longevity of our planet is the extrapolation of the change in oxygen. Contrary to popular thought, the lethal percent of $CO_2$ is about 6 percent, which is 150 times the current level. (*At What concentration Does CO2 Become Toxic to Humans? Principia Scientific International*, Darko Butina, June 27, 2013). So one calculation concerning the life of our planet is when air pollution will reduce or increase $O_2$, or when it will increase $CO_2$ beyond survival limits.

But, what about the second sensitive part of our planet, which affects its longevity? The land portion of our planet contains virtually all of its non-gaseous elements, and that quantity is fixed. Except for molecular changes caused by pressure, or heat, the elemental and molecular structure of our planet is fixed. Even under the unscientific theory of evolution, if these elemental quantities just somehow *naturally evolved* from nothing, into the present residuals, that evolution caused a limitation that has pre-set our capacity to extract these elements from the land for beneficial use. Evolution doesn't allow for the production of more inorganics. Once these elements are used, except for the possibility of re-use or recycle, they are forever lost to future generations. **So we must conclude that, at some point in the future, all extractable natural resources will be depleted. Beyond that point, we have no further advancement.**

The Age of Computers has excited us so much that some of our more emotional "visionaries" are predicting that our everyday lives, and even our very existence, will be completely dependent on computers at some point in the future. Unfortunately, computers, like any natural product, consist of materials, elements, and compounds which are extracted from the land; and afterward are no longer available to build more computers. The *US Geological Survey fact sheet 060-01-USGS* publication lists 20 commonly used metals in computers, several of which are rare and/or expensive to retrieve. Other sources list more than 50 total elements. It takes approximately 1.8 tons of raw materials to make a desktop computer, plus 3.5 pounds of fossil fuel per chip (one or more per computer), plus 400 gallons of water.

Unlike renewable resources, each product, each machine, in its manufacture, has at least one element which is becoming critically more rare and expensive, and will eventually run out.

Industry experts predict there will be more than 64 billion internet devices worldwide by 2025. Nearly 130 new devices connect to the internet every second. The average consumer owns at least four – from GPS-enabled cars to fitness trackers, to home electronics, and of course, smartphones (*Solutions*, Vanderbilt School of Engineering, *One portal, endless possibilities*, Janes Sztipanovits, 2019-2020). The cost to our limited natural resources of this exponential growth in cyber-physical systems, as far as I can see, is not even being considered today. If we extrapolate this idea to our original question, apparently all *natural* solutions to preserving our planet, are futile, and the only permanent solution is supernatural.

## What is the Potential for Saving Our Planet?

We have previously discussed the limitations of our minds. Now we must address its capability in saving our planet. If we examine natural reality as a whole, we must admit that our corporate minds and documentation consists of a finite number of environmental facts, probably mostly undiscovered, or un-accessed. The previously discovered facts, if they are to be complete, must include every sensory perception that every person since the beginning of time has experienced, or the list of environmental facts will be incomplete. This task is obviously impossible to accomplish, since most people have died, and even those still alive have undocumented facts hidden in their brains that they are not even aware of. An example is that primitive peoples are typically closer to the environment and logically understand more about it. Not the why, but the what. Their thoughts may or may not be facts. But over the years, their culture has garnered much truth about their surroundings.

**In this country, we don't even appreciate the wisdom of those in our agricultural societies in the understanding, or at least the recognition, of how the environment works.**

These local observations can open us up to understanding new scientific facts, as many native cultures have helped us develop new natural medicinal remedies. So, in the future, as new facts are discovered and documented, they can potentially be retained in a storage system such as an artificial brain, a computer, or whatever the next generation of storage might be.

The first step in that task must be to retrieve as much of the information in the form of facts as possible. The only source of that information from the deceased is recorded documentation, either written, or otherwise recorded. Because of the exponential increase in the communication of knowledge over time, a significant percentage of additional total information is available from living persons. Each death is a loss of a huge amount of valuable data. So, is there a means of retrieval? Currently, no, and if there is, governments would be justly accused of an invasion of privacy. Therefore the only reasonable solution seems to be to incentivize the population to cooperate in this effort, possibly through census questions.

For the purposes of this book, the best way for the human race to increase the knowledge needed to save our planet is through education to increase the flow of communication between our ideas and our research, and placing the portion of our thinking population in a position to implement decisions that can serve this purpose. One potential step in this process is to analyze the source of environmental knowledge in our society. Most will agree that the more experiences a person's mind has had, the more valuable wisdom therein contained will be. Therefore, theoretically, the older a person experienced in this specialty is, the more facts can potentially be gathered from that source. That is undoubtedly the reason that in the past, old age meant more wisdom. That resource has been largely lost in recent generations because of the view that younger people are more able to adjust with the times and produce greater scientific advances. That perceived capability of the young is undoubtedly due to their more adaptable access to historical information. Too many older people in our society are relegated to retirement or feel irrelevant so that they are not encouraged to participate in decision making.

The key to the value of those facts is their source. Some people are exposed to more significant facts from education and experience than others, but all are of possible value to society, especially in a field such as the environment, where everyone participates. The limitation is that not all experiences are "facts." We hear and read much that is false. One purpose of this book is to urge each reader to establish a personal process based on critical thinking, which would enable the discernment between truth and lies, between science and emotion, concerning the environment.

As we have discussed, the decisions by individuals to take actions, which in the aggregate can save our planet, are made every day by every person on the planet. That challenge then is a matter of adding "environmental awareness" to all educational curricula, and to assure that environmental awareness is limited to science, rather than emotion.

Emotional environmentalism, some say, began in 1859 with Charles Darwin's *The Origin of Species,* the mistakes in which are well known to scientists today. This conclusion is presented in the author's *The Future of Species, The Fantasy of Evolution and the Science of Creation*, Create Space (An Amazon Company), 2015. But the general population is mistakenly taught in public educational systems that microevolution is a scientific fact when, in reality, it is a disproven theory. Beginning in the 1960s, many articles and books were published on environmental issues. In the United States, the Environmental Protection Agency was formed in 1970 from regulations promulgated in the 1960s by the U.S Public Health Service. As happens quite often, the interest in a scientific subject was taken over by non-scientific politicians and media and resulted in an explosion of environmental emotionalism. This trend was popularized by books such as *The Origin of Species*, *Silent Spring*, Rachel Carlson, 1962, and *An Inconvenient Truth*, Al Gore, 2006.

Opening ourselves up to all reality, and therefore, all possibilities, the ultimate solution to prolonging the life of our planet may be up to us, but the ultimate saving of our planet is whether God wills it. He may want it recreated to its original perfection, or He may want it to wind down through entropy into an uninhabitable, sterile planet like the rest of the universe, and be replaced by a perfect eternal Heaven.

Since we do not know His will, or the answer to that ultimate question, the only available way to proceed is to do our very best to use our resources, limited or not, to save the planet over which we have dominion.

**We do not have the right to allow things in our possession to deteriorate into worthlessness, to "go to seed." Otherwise, sustainability and our very existence will be lost.**

*Chapter 16*

# The Spiritual Future of our Planet

## Introduction

This chapter approaches the prophecy of a spiritual heaven from a Christian perspective. The former "skeptic" author of this book, as a scientist, teacher, consultant, and Christian, has spent over forty-five years researching the claims of science and Christianity, some of which was spent attempting to disprove the uncomfortable parts of Scripture. But I failed miserably in that later attempt, and now passionately believe what I will present in this chapter, objectively and rationally. I will not attempt to present herein a detailed study of Heaven, but a summary of the contrasts of the Bible's view of the prophecy of a spiritual Heaven as well as a Heaven on earth, and the secular world's view of a spiritual nothingness. As posited in Chapter 1, at some point, each thinking person must face the question of the possibility of a spiritual reality outside of, and in control of our more familiar natural reality.

## Why Quote the Bible When We are Studying the Mind and Environment?

Most of the readers of this book, as well as overwhelmingly, most of the people in

The United States, claim Christianity as their religion. Therefore this book cannot be objective without addressing those people and those beliefs.

There is a major tendency in the Christian church today to discount certain passages of Scripture as poetic or metaphoric rather than historical. The Bible itself claims that it is completely true, inerrant, and infallible. The following section is partially taken from *A Christian Environmentalist*, E. Roberts Alley, Xulon Press, 2013.

## The Truth of the Bible

The Bible itself presupposes a belief in the inerrancy of Scripture. Deuteronomy 12:32 states, *"Everything I command you, you shall be careful to do; you shall not add to or take from it."* Matthew. 5:18, *"For truly, I say to you, until heaven and earth pass away, not an iota, nor a dot, will pass from the Law until all is accomplished."* Romans 15:4, *For whatever was written in former days was written for our instruction, that through endurance and through the encouragement of the Scriptures, we might have hope."* 2 Peter. 1:20,21, *"Knowing this first of all, that no prophesy of Scripture comes from someone's own interpretation. For no prophesy was ever produced by the will of man, but men spoke from God as they were carried along by the Holy Spirit."* Matthew 26:13, Mark 14:9, *"Truly I say to you, wherever this Gospel is proclaimed in the whole world, what she has done will also be told in memory of her."*

Jesus knew what was to be written in the New Testament. In 2 Peter 3:15-16, *"Just as our beloved brother Paul also wrote to you, according to the wisdom given him, as he does in all his letters as he speaks in them of these matters. There are some things in them that are hard to understand, which the ignorant and unstable twist to their own destruction, as they do the other scriptures."* Indeed Paul tells us in 2 Corinthians 4:2, *"But we have renounced disgraceful, underhanded ways. We refuse to practice cunning or to tamper with God's Word."* Revelation 22: 18-19, *"I warn everyone who hears the prophesy of this book: if anyone adds to them, God will add to him the plagues described in this book."*

These verses speak to the inerrancy and infallibility of Scripture as inspired by God. For those readers who have difficulty believing that every word of the Bible is completely true, even though these verses support that belief (especially Matt. 5:18 concerning the Old Testament and 2 Pet. 3:16 concerning the New Testament), the following comments are offered:

- Virtually all organizations have a charter, a mission statement, by-laws, or some sort of written documentation to describe their purpose. An organization which ignores or changes these standards becomes a different organization from the original. It is certainly reasonable, and in some cases best, to change founding documents when those documents provide the mechanism for amendment. A government can acquire a different purpose, or better provide for its constituents; a company can improve or expand its service, products, or profits. A religion can adapt to the times and attract more followers, but if a Christian Church ignores or changes the Bible, even in the slightest way, it has placed itself onto a slippery slope. If left unchecked, the problem will ultimately cause the church to lose its identity as a Christian Church. It has lost its complete founding document, which was not written by founders but was written by the supernatural Founder. His document, His Bible, allows no provision for amendment, as explained above. Therefore, a Christian Church which allows interpretation of the Holy Scriptures by man, instead of the Holy Spirit, has become a non-Christian sect, with man as its head, in place of God.

- **The Bible must be understood in context, i.e., the Bible interprets the Bible**. I have found that all of the popular so-called discrepancies in the Bible are taken out of context. For instance, in the Old Testament, there are books of history such as the Pentateuch (Genesis, Exodus, Leviticus, Numbers, Deuteronomy), then Joshua through Job. Then, books of poetry and wise sayings such as Psalms, Proverbs, Ecclesiastes, and Song of Solomon, and books of prophecy, such as Isaiah through Malachi. The purpose of the books of history is to communicate historical facts, not symbolism or literary devices. The purpose of the books containing poetry and wise sayings could be seemingly contradictory since they consider a wide variety of situations, some with opposing solutions. The truth for the prophetic books must use some symbolism, since they refer to the spiritual world as well as events in the future.

- Likewise, in the New Testament, there are books of history, such as the Gospels (Matthew, Mark, Luke, and John), then The Acts of the Apostles, letters such as Romans through Jude, and the prophetic book of Revelation.

- Most supposed contradictions come from either Proverbs or the Gospels. Proverbs, in context, is a book of wise and true sayings by Solomon, which, as in life today, teaches one action or response in one case, and perhaps an opposite action or response in another case – all being true. The Gospels are written by observers who witnessed, or researched (Luke) and emphasized different and similar events.

- By virtually all definitions, God is supernatural, supreme, the creator, omniscient (has infinite knowledge), omnipresent (present everywhere at all times), omnipotent (has unlimited power), and omnibenevolent (has perfect goodness).

- A perfect God logically must do everything perfectly, or He wouldn't be God. This includes communication with His creation. He has elected to do that in the written form, the most advanced technology of communication available at the time of the writings. The Hebrew people were appointed by God to copy the manuscripts as originally authored. Romans 3:1-2, *"Then what advantage has the Jew? Or what is the value of circumcision? Much in every way. To begin with, the Jews were entrusted with the oracles of God."* By law, these scribes transcribed these manuscripts perfectly, or the manuscripts were destroyed. The success of this practice has been confirmed by comparing transcribed manuscripts discovered many centuries ago with those which were transcribed even earlier but not discovered until 1947 (the Dead Sea Scrolls). Before the discovery of the Dead Sea Scrolls, the earliest known manuscripts of the Old Testament were not written in Hebrew, but copies of the Greek Septuagint, which was translated from 350-140BC. (*The Story of the Bible,* Larry Stone, Thomas Nelson, Nashville, TN, 2010). God would be out of character if He were to communicate imperfectly or allow His Word to be translated with doctrinal errors or mistakes. He is not a God of confusion.

- The Codex Vaticanus was the first combined manuscript of the Old and New Testaments and was written in Greek in 300-350 AD. It was not discovered until 1475 AD, is the oldest existing copy of both testaments, and includes the complete Septuagint.

- Christians, and Jews accepting Jesus Christ as their Savior, as the new Israel, are now responsible for preserving God's Holy Word. 1 Corinthians 4:1, *"This is how one should regard us, as servants of Christ and stewards of the mysteries of God."* Romans 3:29-30, *"Or is God the God of Jews only? Is He not the God of Gentiles also? Yes of Gentiles also, since God is one. He will justify the circumcised by faith and the uncircumcised through faith."* Romans 9:6-8, *"But it is not as though the Word of God has failed. For not all who are descended from Israel belong to Israel, and not all are children of Abraham because they are his offspring,*

> but *'Through Isaac shall your offspring be named'. This means that it is not the children of the flesh who are the children of God, but the children of promise are counted as offspring."* (See also Hosea 2:23, Romans 9:24, 10:12; 11:25-27; 15:8-21, and Galatians 4:28.)

In this book, as in most literature, the word *Scripture* is defined as *a body of writings considered sacred or authoritative* (Webster's Ninth New Collegiate Dictionary); and to Christians, as *the books of the Bible*. Scripture is not natural. It does not claim to proceed from the memories and minds of its writers and scribes. Rather, Scripture is all supernatural, having proceeded from God through His children on earth.

If Scripture is indeed supernatural, it overrules the natural and must be accepted, even when it seems contradictory to the natural. This belief statement is taken from Scripture itself, as quoted above.

If we count any word of Scripture as being false, exaggerated, or only a literary device, we have the logical right to discount any, or all, of the Bible; losing the standards of Christianity. Then our religion would have no basis, no character, no "by laws." We can argue that men made mistakes in transcribing the original Scriptural manuscripts. But then, we limit the communication of a perfect God. He would have failed in His goal of explaining Himself to His Creation. We can likewise argue that the original manuscripts may have been true, but the translators misinterpreted God's directions. But we must then logically admit that God may have been originally accurate in His communication, but failed in its preservation. God is either God by definition – omnipotent and omniscient – or He is not really God. If He is truly God, He makes no mistakes in communicating His Word orally or in written form.

Another argument against the inerrancy of Scripture is that the Greek manuscripts were slightly different in the Alexandrian Greek family of manuscripts written about 300 BC and discovered in 1896, AD, and the Byzantine family of manuscripts, written about

1550 AD. I have compared the Greek words of these texts and found no significant doctrinal differences. Mostly minor conflicts occurred when the scribes of one family of manuscripts added to or left out words or phrases. Other differences are due to the fact that English is a much weaker language than Hebrew or Greek, and has many fewer words than either original language, so it is difficult to translate without some paraphrase.

The proof of the authenticity of Scripture is the fact that over 5500 separate parts of ancient handwritten manuscripts have been discovered, and none contain significant doctrinal or historical conflict. This is many more manuscripts, with much fewer inconsistencies than any other book in history, including the writings of Shakespeare.

There are numerous examples of the tendency that man has to edit God's Word. That was not allowed as long as Jews preserved the Word. But we Catholics, and especially Protestants, have not been as faithful:

- The original New Testament canon was not accepted as canon by the early Church until around 300 AD. Frederic Kenyon, as quoted in *The Story of the Bible,* Larry Stone, Thomas Nelson, 2010, states, "*We have a period of rather over 200 years (after Revelation was written) when the various books circulated... with no central control to ensure a uniform text...Christianity was exposed to persecutions by the Roman Emperors and Governors, when copies of the Scriptures were a special object of search and destruction... [Christians] were thinking only of the substance of the Christian teaching, and caring little for the verbal accuracy of the text.*"

- One criteria of canonization was that a New Testament book had to be written by one of the twelve Apostles or Paul, Jesus' brothers, James and Jude, or in the case of Mark and Luke, with apostolic approval. The author of Hebrews is unknown. The Old Testament canon had been effectively established before the time of Christ. There were several attempts at canonization in the early AD, but the early Christian Church did not accept the New Testament books because they were included in a canon, but because they were already regarded as divinely inspired. The most important and influential translation of the New Testament was contained in the Latin translation of the entire Bible in the early fifth century, called the Vulgate. It was the official translation used by the Roman Church for more than 1500 years (ibid). There have been other rather insignificant disagreements about the canon of the New Testament Scripture over the years since the Reformation.

- The early Protestant Church, placed Hebrews, James, Jude and Revelation as a supplement to the New Testament, without listing in the Table of Contents (*The General Epistle of James*, Eerdmans Publishing, Grand Rapids, MI, 1976). This mistake was probably because of the influence of Martin Luther, who arbitrarily, and without the support of ancient manuscripts, proposed this change in the older church canon.

The order of the New Testament did not conform to the books of the Vulgate until the publication of the Great Bible in 1539. Martin Luther said that the Saint James Epistle *has no gospel character to it*. Erasmus, in 1516, said that James lacked apostleistic gravity. Tyndale, in 1534, followed Luther and omitted these four books from the New Testament Table of Contents, as did Calvin in 1551, but they did not deny that they were Holy Scripture.

Biblical criticism, with the excuse of avoiding dogma and bias, became trendy between the Reformation and the Enlightenment and has attempted to reconstruct history according to contemporary understanding. It has lasted off and on to the present and has become a source of confusion and schism, rather than comfort and peace, as the Bible originally afforded. Many Protestant

Church splits have been caused by these radical and unnecessary interpretations.

One of the most successful of these criticisms today is that the books of the Bible were written to the readers at the time, and are not applicable, or as applicable, at present, due to our different circumstances. But if we consider the Bible as a spiritual book, as I have suggested, this criticism is invalid since, in the spiritual realm, there is no time (Ps. 90:4: *For a thousand years in your sight are but as yesterday when it is past, or as a watch in the night.* II Peter 3:8: *But do not overlook this one fact, beloved, that with the Lord one day is as a thousand years, and a thousand years as one day).* The truth prevails throughout history and is not changed or altered by time, or other natural circumstances, trends or beliefs.

R.V.G. Tasker in 1958 said in his *The Second Epistle of Paul to the Corinthians, Tyndale New Testament Commentaries,* Eerdmans Publishing, Grand Rapids, MT,

*Nowhere in the New Testament is heaven described; but in the poetical imagery of the Revelation of John are to be found flashes of its glory sufficient to stimulate the imagination of the saints, to encourage them in their sufferings upon earth, and to intensify their longings for their permanent home.*

A very weak comment; how else can we experience the spiritual, other than through visions? That is God's method of allowing John and us to better understand the spiritual. The one who had the vision was not just someone dreaming; he was the Apostle whom Jesus loved who saw the visions "in the Spirit". We cannot experience the spiritual through education, since there is no one alive who has experienced all of the spiritual. We have felt the spiritual and heard the spiritual and become one with God in the spiritual realm, but we haven't experienced the total greater truth. So we must depend only on God so that we can draw closer to Him, to be one with Him, to understand Him. The book of Revelation is not poetical imagery, but a true recorded vision of all God wants us to know about our glorious future. We don't need our imagination to understand these truths of the reality of Heaven; we just need to accept the Apostle's vision as we accept the words of the rest of the Bible; from God to us in truth and love. The book itself makes this very clear: John testifies that he heard and saw these visions (22:8), that the trustworthy and true words were communicated to John by an angel (22:6), sent by Jesus (22:16), never to be sealed up (22:10), to bless those who keep His Words (22:7), and to curse those who add or take away from the words (22:18).

Some commentators have assumed that this blessing and curse refers just to the Book of Revelation, but most understand that they refer to the entire Bible. The Westminster Confession of Faith also takes this stance. Throughout the Bible, this is confirmed as "The Bible interprets the Bible." See Deuteronomy 4:2, 27:26, and Galatians 1:8, for example.

Perhaps the most flagrant and dangerous abuse of spiritual acceptance is the rather recent belief in the Literary Framework view: *a literary devise to portray the creation week as if it were a workweek, but without concern for temporal sequence* (*Genesis and Science, Introduction to Genesis,* Desmond Alexander, *English Standard Version, Study Bible,* Crossway Bibles, Wheaton, IL, 2008). The section containing this reference, *Genesis and Science,* demonstrates the misunderstanding of certain theologians, of the spiritual nature of God. There is no need, and as Christians, we have no right to question God's words, or place those from a historical book like Genesis in a category of myths, legends, or a literary device. Alexander goes on in this and subsequent sections to dig himself deeper into heresy. Authoritative texts like this have presumably influenced contemporary theologians to believe that science has the right, and even the duty, to speak to theological truths, when, as I have mentioned previously, scientists typically have no education or training in spiritual matters. Science is subservient to the spiritual reality, not vice versa. This issue is discussed in detail in *The Future of Species,* by the author, from a scientific perspective as well as a theological perspective. I can testify, as an environmental scientist, factually, that the order of creation, as described in Genesis, matches the latest scientific evidence, especially in DNA research, much more closely than any evolutionary theory. Also, God's Word states that the Sun was not created until the fourth day. Before that, there was no way to measure time on the earth. Therefore, the first three days could have lasted for millions or billions of years or could have lasted one second. So there is no need to play God by interpreting the words of Scripture as subject to science. God, and God alone, created science, and it is our responsibility to use science to understand natural reality, and not to waste time delving into spiritual things, unless we are indwelt by the Holy Spirit, and then God can enable our spiritual nature to truly understand ultimate truth. Only then will we really be able to save our planet

## Heaven as our Planet's Future

The advantage logically of accepting the belief of a spiritual reality is that it is the only rational explanation for the source of natural reality. No one has proposed seriously that the Big Bang came from nothing, since virtually all intellectuals admit that a negative belief is no belief at all. In other words, a belief that something does not exist is unscientific and nonobjective. We can't base a belief on a negative that is impossible to prove. If we could, there would be no science.

The existence of a spiritual reality is the only explanation for the complexity and the continuity of natural reality. This is true from outer space to the smallest discovered matter, neutrinos. It is also obvious when we examine our DNA, stem cells, vision, lungs, heart, brain, and mind. Chance and the survival of the fittest cannot explain these miracles. Even Charles Darwin believed in a Creator God (*The Origin of Species*, Charles Darwin).

So, once we accept that a spiritual or supernatural reality must exist, along with a natural, we must ask how we fit into that dichotomy. No scientific answer is possible, since the question is beyond scientific knowledge. No non-Christian religion has invented an explanation as simple and logical as is presented in the Christian Bible. All other theories are dependent on human obedience and works, rather than completely on the grace and love of a supreme supernatural being we call God.

Simply put, the Bible teaches that: 1) The Trinity, God, Jesus and the Holy Spirit, has existed in the spiritual reality forever. 2) When man was created with a mind and free will to use it, he decided that he wanted to take over the natural reality and replace God. 3) Because of that misguided arrogance and selfishness, man was punished by not living forever with God, as was the original intent; and instead, along with all other living organisms, was cursed with death and enmity with nature. 4) God's love sent His Son to earth to live, teach, and primarily be punished for man's sin of wanting to take God's place. 5) With that punishment, humans who, called by God to use their free will, accept this punishment and sacrifice of God's Son Jesus Christ, can once again walk with and be associated with a holy God on this earth. 6) When natural death comes to humans, those who have accepted Jesus' sacrifice for their sins, are transferred from the natural reality to the supernatural spiritual reality to exist forever with God in Heaven, as was originally intended. Those who decide not to accept this sacrifice are transferred from the natural reality to the supernatural reality to exist forever under permanent torment with Satan, who originated the idea of taking God's place.

## What the Bible Says About Heaven

The Book of Revelation gives a description of heaven, which is both enthralling, and at first, confusing. But how else can something completely spiritual be communicated to one who has not experienced a spiritual realm? The last two Chapters, 21 and 22 of Revelation describe the following (a caution is that since the entire Book of Revelation is a description of the Apostle John's vision, much of the language must be symbolic in order to explain the supernatural in terms of the natural; but even though symbolic, every word is true. In other words, the symbolism is an accurate depiction of Heaven as John saw it, and as we will experience it.):

**21:1**: The earth has passed away, replaced by a "new Heaven and a new Earth" (not evolved, but new). When God makes our new Heaven, He won't just modify the old earth, He will destroy and replace it. Likewise, when we become Christians, God won't just modify our old selves, He will destroy them and replace them with a brand new self. Other references are Job 14:12, Ps. 102:26, Is. 34:14, 51:6, 65:17, 66:22, Matt. 5:18, 24:35, Heb. 1:11, II Pet. 3:10-13.

**21:2**: Heaven, the New Jerusalem, is the Bride of Christ, and takes the place of the natural Bride, the church.

**21:3, 4,22:** God and Jesus will be with Their children in the temple or tabernacle, in place of a structure. There is no temple in the City, for the temple is *the Lord God the Almighty and the Lamb*. It replaces the church (see also Gal. 4:26, Heb. 12:22, 23).

God will wipe away every tear, mourning, crying, pain, and death. See also Luke 20:36. There will be a fullness of joy and pleasure forever (Ps. 16: 11). There will be no marriage, perhaps no gender.

Isaiah 57:15 describes this existence as follows:

> *For thus says the One Who is high and lifted up, Who inhabits eternity, Whose name is Holy: "I dwell in the high and holy place, and also with him who is of a contrite and lowly spirit, to revive the spirit of the lowly, and to revive the heart of the contrite."*

Matthew 5:3,10 says, "Heaven is for the *poor in spirit*, and *for those who are persecuted for righteousness sake."*

**21:6:** God will provide all spiritual needs without cost.

**21:7:** All believers will be sons of God. See also Luke 20:36.

**21:8:** All cowards, unbelievers, abominables, murderers, immoral persons, sorcerers, idolaters, and liars will die into the other supernatural option, hell.

**21:10-21:** Heaven is immense, 1380 miles (10,000 stadia) cubical. The height gets into the ecosphere, where, in our universe, there is virtually no oxygen, just hydrogen and helium (It is 32°F at night and 3000°F during the day). The city appears like clear gold and has twelve gates inscribed with the names of the twelve tribes of Israel. The jasper wall between the gates has twelve foundations inscribed with the names of Jesus' twelve apostles. It is 216 feet (144 cubits) thick. All these dimensions are multiples of 12 or 1000.

**21:22,23:** There will be no sun or moon; the glory of God gives it light, and its lamp is Jesus. There may consequently be no gravity and no mass, but there may be; God can do whatever he wants, especially in an existence beyond science. There will, therefore, be no time, apparently no aging, no different ages. Everyone will presumably be the same age, as God's equal children.

Genesis 1: 1-7: God originally created an earth with no death, perhaps a preview of what Heaven will be like, *In the beginning God (Hebrew, singular) created (Hebrew "He created," plural) the heavens and the earth. And the earth was formless and void (a waste and emptiness), and darkness was over the surface (face) of the deep; and the Spirit of God was moving (hovering) over the surface (face) of the waters. Then God said "Let there be light," and there was light. And God saw that the light was good; and God separated the light from the darkness. And God called the light day (yom), and the darkness He called night. And there was evening and there was morning, one day.*

The first created earth was formless and void; no shape, empty; no gravity because of no sun, moon or stars (Gen. 1:12). No gravity, no mass, no sphere, no spinning earth, no circuit; just an unshaped mass of water.

*Genesis 1:6-7: Then God said, "Let there be an expanse (firmament) in the midst of the waters, and let it separate the waters from the waters. And God made the expanse (firmament) and separated the waters which were below the expanse (firmament) from the waters which were above the expanse (firmament); and it was so."*

At the end of this second day, there was still no dry land (day 3) and no universe (day 4), therefore there was still nothing to form the earth into a sphere with a fixed shape. There is a possibility that God is telling us here that the new Heaven will be similar. We know, as described above, that there will be no universe, so there could be no gravity or mass at that time to hold the earth together. But God can, and undoubtedly will see to it that John's vision is true, and we will enjoy a perfect environment forever in a much better way than our universe has provided.

**21:24, 22:5**: There will be no night, perhaps no need for sleep.

**21:25:** Its gates will never be shut.

**21:27:** Only those whose names are written in Jesus' Book of Life will be in the City.

**22:1, 2:** The River of the Water of Life flows from God's throne through the middle of the clear golden street. The Tree of Life, yielding twelve crops of fruit is on either side of the River. The leaves of the tree are for the healing of the nations.

**22:3:** God and Jesus' throne is in the city.

**22:14:** God's children have the right to the Tree of Life.

## The Kingdom of God and the Kingdom of Heaven

Theologians differ in their interpretation of these two terms. Probably the best summary is in *Naves Topical Bible*. The Moody Bible Institute of Chicago, Revised 1974; but it is based only on the King James Version with "liberal uses of the words in the Revised Version."

The *Kingdom of God* is mentioned in Mark, Luke, John, and throughout the New Testament, but the expression *Kingdom Of Heaven*, is only in Matthew. Both the Greek Byzantine and Alexandrian texts include verses in these books that indicate that the *Kingdom of God* is a spiritual kingdom in heaven, rather than a Church Universal entered on the earth, as is the *Kingdom of Heaven*. This earthly church, which Jesus left for us in His stead, is inhabited by God's children on earth, before their natural death, who will then be transformed into new beings, just as the universe is transformed into Heaven upon Jesus' second coming. But this earthly Church is also inhabited by those who are not God's children, just as it was in the time of the writing of the New Testament books.

According to *Nave's*, the Bible speaks of the *Kingdom of God* and the *Kingdom of Heaven* with each having different characteristics as follows, from the King James translation:

|  | **Kingdom of God** | **Kingdom of Heaven** |
|---|---|---|
| **Entrance** | By a new birth | By manifold righteousness |
|  | John 3:35 | Matthew 5:20, 7:21 |
| **Future of souls** | Secure | May be cast out |
|  | John 3:18, 5:24, 10:28 | Matt. 8:12, 13:41,42,47-50 |
|  | II Timothy 4:18, James 2:5 | 24:50, 51, 25:30 |
| **Existence** | Eternal, no death | Comes to an end |
|  | Daniel 4:2,3, Hebrews 1:8 | I Cor. 15:24, Revelation 20:6 |

|              |                    | Luke 20:36              |
|--------------|--------------------|-------------------------|
| **Inhabitants** | Only the saved  | The saved and the unsaved |
|              | John 3:3,5, I Thess. 2:12,13 | Matt. 13:37-43, 47-50 |
| **Marriage** | None               | Optional                |
|              | Matt. 22:30, Luke 20:35 | I Cor. 7: 8-16      |

In summary, according to *Nave's* interpretation, the *Kingdom of God* is New Jerusalem; what we call Heaven, and the *Kingdom of Heaven* is God's created kingdom on earth, which Jesus left for His children as His universal Church. The Apostle Peter in II Peter 1:11, uses the term: *everlasting kingdom of our Lord and Savior Jesus Christ*.

Some who believe that the reason the *Kingdom of God* is not used in Matthew is because the word God, *Elohim,* could not be spoken in Old Testament times. But the word *God* is written hundreds of times in the Old Testament. A good example is Leviticus 19, where *I am the Lord your God* is used fifteen times with different Hebrew words for *God and Lord*. God is also written many times, especially by the Prophets, as *Yehovah*. So, we should understand that there is a temporary spiritual Heaven on earth, the Church with all its flaws, being managed by sinners; and there is a permanent spiritual Heaven, which will replace the church, and be ruled by a just and gracious God.

Those who believe that the two terms are interchangeable also have a good argument that God's church is spiritual, just as is Heaven, but they must admit a huge and critical difference.

Therefore, as we work to save our planet's resources and quality, we should admit that the future of this present universe is only temporary, and prepare ourselves for an ultimate and permanent spiritual existence. That preparation can realistically and practically be accomplished by God, the Holy Spirit, through our reading, studying, learning, and following the directions in the Bible, the only book ever published, claiming to be the direct communication between God and His Creation. Those directions basically say what one would expect, that we must give up our claim to the right to ourselves and accept Jesus' sacrifice for forgiveness, which enables us to be holy enough to have relations with a holy God.

Given this description of our spiritual future, the question remains: Is the scientific mind capable of saving our planet? We have seen that if we can cease the futile Culture War, the scientific mind is potentially capable of making our planet more inhabitable by preventing the destruction of our air, water, and land quality through pollution, and by partially preserving the created natural resources of the planet. Of those two goals, environmental protection is very expensive, but theoretically possible, but complete resource preservation is impossible, because of the continued necessity of the use of energy by humans. Once coal, oil, gas, and nuclear fuel are depleted, they cannot be recovered. The only known recoverable energy sources are solar, wind, water movement, geothermal, hydrogen, and vegetative fuel, and we have discussed each of them. The fact is that they are either prohibitively expensive, at the present time, or they are not available all of the time and at every location. They will help our planet last longer, but they won't solve the ultimate deterioration.

Our only ultimate solution is to accept God's Word that at the proper time, known only to Him, this planet will be replaced, not renewed or evolved, by a new spiritual existence called Heaven, as the Kingdom of God. We are not given enough information about Heaven to accurately visualize it, nor should we. But we do know that people and things may appear as they did on the earth, but they will not be natural, or subject to the rules or limitations of science or nature.

We have seen that there will be no sun or moon, therefore, there will be no universe, no spinning, spherical earth, no time, no gravity, perhaps no mass. But we have also seen that there will be water, trees, fruit, structures, and precious minerals. Our comfort should be that, regardless of a redesigned reality, there will be no death, no lies, no class distinctions, no marriage, perhaps no gender, no sadness. There will simply be a permanent existence with God, in enjoyment of peace and love. So we cannot even compare the advantages of our current existence with our future existence as children of God.

So the answer to the question of this book is YES for our planet and NO for our permanent existence. The scientific mind can increase our comfort and longevity, and allow for the sustainable growth of our civilization for only a limited time. But God has ordained that this planet will come to an end, and be replaced by something much more suited to our ultimate existence.

What can we do then to assure this future for ourselves, our families, and our world? There is only one answer; to first, and temporarily, research, understand, and submit to a spiritual truth, even a spiritual existence on this earth. This existence, called the Kingdom of Heaven described above, carries a serious responsibility to do all within our power to protect our resources and environment, because we do not know when this temporary existence will come to an end.

Our second responsibility is to assure the permanent and ultimate future of our existence by accepting its spiritual nature and the free offer to be a part of this wonderful future. The Bible is filled with descriptions of how we can accept that free invitation to be a permanent child of God; to come to Him as a child. The *foolishness* of the world as a child, is many times accompanied by the *wisdom* of the Spirit. Sin and even science lead to complexity; God's truth leads to simplicity. That God-given wisdom of a child will enable us to give up ourselves in submission to the truth of Jesus Christ. One of the simplest explanations of this is in Revelation 3:20, when Jesus says, *"Behold, I stand at the door and knock. If anyone hears My voice and opens the door, I will come in to him and eat with him, and he with Me."* This is an invitation directly from God requesting a response from us. We can RSVP by opening the door as an eager and trusting child, and accept the Lord's Supper to dine directly with Jesus and He with us, forever and forever on earth and in Heaven, or we can keep our door closed. By doing this, we act as a selfish child, hiding in his room, and denying Him His invitation and thereby, follow our only other supernatural choice, Satan. No dining then; only continuing punishment exceeding the worst we have experienced on the earth, forever and forever in hell. Our choice is possible through the gift of free will God so graciously gave us, and the gift of forgiveness He also gave us through the sacrifice of his Son Jesus in complete payment for our sin of wanting to take his place.

# BOOK VI:
# A CHALLENGE TO SUCCEED

*Chapter 17*

# Conclusion: Our Natural Planet Can be Saved

## Introduction

We have examined in some detail, in Chapters 1 through 8, the brain and the mind and their place and use in the environment. We have seen the possibilities of using these tools to resolve natural and secular problems if we will release our obsession with participating in the Culture War. In Chapters 8 through 11, we examined in detail the constituents, actions, and reactions of our environment. Chapter 12 introduced the often overlooked crisis of the preservation of our limited natural resources. Chapters 13 and 14 offer suggestions as to what we can accomplish as a society and as individuals, and Chapters 15 and 16 discuss the future consequences of our decisions. The conclusion of this discussion is that life on this planet will end at some point in the not too distant future, if we refuse to limit the discharge of the residuals of our civilization into the environment, and if we refuse to limit, return and recycle the limited natural resources of our environment in our quest for more and more technological gadgets and machines.

## The Challenge

The challenge to us today as a society is to admit that reality consists of both the natural and spiritual and to turn, or advance, to the acceptance that we must thoroughly understand the natural, and submit to the spiritual, to have any hope of saving our future. Our scientific forefathers realized this challenge through the first part of the Enlightenment, as discussed in Chapter 2, but we have lost that vision, as we have attempted to advance as if we were God Himself, without submission and understanding of His Creation and its purpose.

In answering this challenge, I urge you to consider the recent scientific advancements of our age. Consider also how much of that progress opens our eyes and removes our blinders, so that we can see and understand the indicators of the presence of God: the ability of DNA to define and program life, the superiority of medicine based on living organisms, and the existence of nanoparticles which have energy and mass, but as yet no discernable purpose. These examples, and many more, prove that natural reality is an indicator of God, as the supernatural source of planning, intelligence, and reason, which together enable life itself. These scientific advances are not only detectors of God Himself, but can be a potential of almost unlimited resources, which can save our planet until God wishes it to be converted into a permanent rest for His children.

## The Societal Pressure to Defeat our Challenge

Our challenge is not an easy one. Most social pressures including the participants in the Culture War, the academia, the liberal establishment, and the already wealthy, are desperately striving to defeat the truth; and because of these pressures, the timid and dependent media, has bought into a socialistic mentality in which the end justifies the means. They have deteriorated in their reasoning process to the point that an immediate "Band-aid" fix is preferable to a slower, more rational fix, in order to receive immediate gratification from their peers and dependents. They prefer the unjust redistribution of wealth to everyone equally, to the encouragement of wealth accumulation and philanthropic, voluntary giving to the poor and unabled, as has been practiced by Christians over the ages. This naive attitude will defeat our progress as a civilized society, as it has historically, every time it has been tried. Wealth accumulation is a positive societal virtue, **as** long as the accumulation is used to advance civilization and the lives of those not gifted with wealth. It is especially valuable to society when it is able to replace government-subsidized "entitlements."

This book has attempted to stick to scientific facts and avoid emotionalism and simple feel-good solutions to complicated environmental problems. This country and this world deserve nothing less.

A person, or a group of people has no right, regardless of their position, power, experience, or education, to *unilaterally* decide what is real, permanent and good for the rest of society. If that were allowed, we would never discover anything unknown at that time. Science would be useless, and we would still have slavery, believe the earth is flat, and know of no particles smaller than an atom. Using courts and politics instead of laboratories is not the answer to civilized growth and survival.

So, insistence without proof is certainly legal, but it is the repository of ignorance and selfishness in the Culture War.

As, hopefully, the readers have seen, this book has two additional broad purposes concerning our natural existence. One is to communicate a concern for the future of our planet to individuals, who collectively can alter their daily living habits and practices in these areas of environmental awareness, to the point that our civilization actually controls its residuals and manages its use of our limited natural resources, in ways that can help us save the planet.

But the second and equally, or even more critical purpose of the book, is to ask what entity in our society is capable, responsible, and motivated enough to coordinate the governmental and private resources in our country and in the world, to begin the thought process necessary to implement change.

If we do nothing more than we have, our civilization will die. We don't seem to even care enough, as we progress along our current and comfortable journey into the future toward destruction, to admit that we must change our priorities to survive.

Now we spend all of our intellectual resources feeling good about our insistence that industrial regulation increases can solve our climate irregularities, or changing the subject by whining about the lack of diversity, medical care, and education in our society.

It is time that we cease our complacency and our efforts to prove that we are tolerant, inclusive, and kind. Those qualities are obviously critical for our survival, but they should be natural and expected. All the efforts towards changing our national mentality in order to force and adopt these qualities, ignore the more critical need of saving our planet.

We can be diverse, provide the tools for survival to the disabled and the poor and needy of our country and the world, but if we end up with a socialistic society where we are all equal in wealth, health, and perceived happiness, who will lead us to the next level of survival? Who will fund that effort? Is it enough to think we are wealthy, healthy, and happy, or is it more important to survive?

Who will fund the effort to survive? Who will provide the education, design, labor, and land needed to implement that effort? What are the possibilities? Who has the motivation, the power and the finances?

Should we just pass this on to the government? They have failed so far to even address the real problems.

**Is the purpose of our government to control society, or to serve society... or both? Or is it to see that society survives?**

Is there any practical entity that has the resources available to proceed with this challenge? And let's face it, the reason government has that potential is their control over their constituents; their taxing power over the citizens. The only other practical possibility is private foundations. But with our current political system, are they motivated to look at the larger and more critical challenge presented in this book, or will they just continue to follow their pet projects in order to seek the recognition for their efforts in art, diversity, health care and education?

So, my conclusion is that our government must accept this challenge to save our planet. And since the United States has a powerful, influential, and wealthy government, it should lead the charge. We must resist our tendency to make the world just like us. We must instead set the example of how we can provide the motivation, the organizations, and the resources to save our planet. We must have more government-owned or government-subsidized research facilities, but these facilities should be involved in research to save the environment and to find replacements and processes to save or replace our limited natural resources, instead of wasting their efforts in prideful, but futile projects, like listening to possible voices from outer space, trying to find signs of water or potential life on other terrestrial bodies, or developing alternative energy sources like solar energy, wind, and waves, which all can help in our survival but none of which, even as an aggregate, can provide our ultimate survival. Let those projects happen as well, but let our primary effort be to stabilize our environment and replace our limited natural resources.

## One Practical Possibility of Financing the Challenge as an Example

Nestled almost in secret in the center of the United States is the Oak Ridge National Laboratory (ORNL), occupying 10,000 acres of the 35,000 acres of the Oak Ridge Department of Energy Reservation, the birthplace of the atomic bomb. There is finally a concerted effort to commercialize the technology developed in Oak Ridge by partnering with American industry. Perhaps that recent emphasis is to ensure the survival of this, the largest of the 10 Department of Energy science labs in the country, but it is still trying to serve the people of this country, as well as the vested interest of the federal government. In October of 2019, ORNL put eight new discoveries on display at the Third Technology Innovation Showcase, for consideration by private companies and entrepreneurs for license agreements for development into commercialization.

This chapter questioned the use of government funds for space exploration to find extraterrestrial life, but there is no question that the federal government is in the best position to perform general research into more critical and valuable future technology, providing that this research is available to the citizens of the country for the advancement of civilization and the preservation of the planet, and not reserved exclusively for the power of politics, private industry, warfare or space exploration. Think of the potential result.

The only practical option to this solution of future technical advancement seems to be subsidizing or otherwise incentivizing private enterprise to be more involved in research for the common good. But a partnership of private enterprise and federal government has far more potential if it is monitored by the Legislative Branch of our government as our representatives.

Today's annual budget for ORNL is $1.4 billion. In May of 2019, Energy Secretary Rick Perry announced that Oak Ridge would have a new exascale computer operable by 2021, which will be the world's fastest computer. The existing supercomputers at ORNL are currently used for new nuclear medicines, energy, transportation, and materials development by many public and private researchers in a variety of fields, including artificial intelligence, bioresearch, transportation, physics, chemistry, and nuclear power.

The Laboratory is also home for the big Area Additive Manufacturing Machine. This additive, or 3D printing machine, is already playing a key development and testing role for new production manufacturing technology at over 700 companies. The Lab Director, Thomas Zacharia, has stated that their goal is to turn scientific research into practical societal gains.

So, our job as citizens is to insist that these facilities, funded by us, are used to advance science and technology to the point that we will be able to preserve our planet for as long as possible. These public facilities should not be the exclusive playthings for the government, for exclusively large corporations, the military, or for any vested interest besides that of our civilization and its survival.

## A Caution

For those of us who accept the reality expressed in Chapter 16, that God will determine our future, we must guard ourselves lest we become desperate to save ourselves as ones who are in complete control of our future. Perhaps our current dire state is not a result of our technological advancement, as we imagine, but a result of our sin, greed, and selfishness; by our wanting to be God, to be in control of our future, just as Adam and Eve. So, yes, we may save our planet for a while; we certainly should, that is our responsibility as stewards; we can take pride in that. Isn't that the goal of our lives; to save ourselves and others? We may indeed extend the life of our planet, but only God can save our civilization!

# Conclusion

In this book's Preface, I asked, "Does our planet need saving?" Is it in order for each of us to do our own will, to narcissistically be what we want to be, selfishly having our own way, regardless of those around us? That is man's view of our future. On the contrary, God's view of our future is a paradise of love, peace, comfort, worship, and enjoyment of God and of His Creation, including others.

If we choose our own view, as opposed to God's view, we will destroy our planet, perhaps not its environment and natural resources, but certainly its true moral fabric, and its civilization. History has well taught us this lesson over and over again, from Babylon through the Roman Empire, from Nazi Germany through Soviet Russia, from empires and dictatorships through all forms of socialism. We can only have a future of true freedom and advancement if we have a society, a planet, of submission to God's Word, regardless of how exclusive that sounds. God's Word, in its context, has never been proven wrong, or weak, or inadequate, or even exclusive, other than in its acceptance. It has survived through all of society's experiments and failures, as a model of future success.

We must then ask the frightening question: "if our will overrides God's will, will God end the life of our planet?" Obviously, our will cannot override God's will; our part of His reality is obedience, rather than trying to influence Him or His plan for us. But since God knows us and our sin nature, will he let us dig our way into destruction? The Book of Revelation seems to indicate that at some time in the future, that will occur; He knows that our tendency is to destroy ourselves.

But when? God has placed us here; therefore, now our job is to be content with that placement (Phil. 4:11, Heb. 13:5), and do all within our power to see that *Thy Kingdom come, Thy will be done on Earth as it is in Heaven* (Matt. 6:10). Our job as stewards of God's creation is not to end the life of our planet, but to preserve it.

God will decide if our seed is to be limited to one generation, or whether many generations will live in the future to enjoy the wonders of the truth of His Creation. Our job is not to play God, and force, or allow, the end of our planet, that's God's decision, and His job exclusively!

While we pursue our best efforts of medical and nanoparticle understanding and use, as well as other beneficial scientific advancements which are being developed, through individual and public/private partnerships, each of us must concurrently face the perhaps uncomfortable truth, that our future is ultimately in God's hands, not ours, and submit to that future of reality as we submit to His direction for taking dominion over the earth, and rest with comfort and peace that he will provide that ultimate future peace.

# About the Author

E. Roberts Alley is a registered Professional Engineer and scientist by education and has practiced and taught environmental engineering for fifty-nine years, working primarily for industrial and municipal clients to lower their residual discharges to the environment.

Bob has BE and MS degrees from Vanderbilt University and was an Associate Professor in the Department of Environmental and Water Resources Engineering at Vanderbilt; and taught postgraduate environmental courses at the University of Tennessee, George Washington University, Jefferson State University, East Tennessee State University, The University of Tennessee at Chattanooga, The School for Continuing Engineering Education, The US Corps of Engineers, The Centre for Management *Technology,* across the US, and in England, France, Singapore, Malaysia, and Indonesia.

He was called to be a Sunday School teacher about forty-five years ago, and since then has served as a Deacon, Elder, Christian School Board founder and member, and a Bible Study teacher. He lives in Nashville, Tennessee and is a member of the Presbyterian Church in America. He has four wonderful children and has been blessed with eleven grandchildren and memories of Marion, his wife, who passed on in 2012.

# Books by E. Roberts Alley

*Routine Quantitative Test for Salmonella Organisms in Natural Waters,* 1973, Vanderbilt University

*Drainage Management,* 1991, University of Tennessee Center for Government Training,

*Air Quality Control Handbook,* 1998, McGraw –Hill

*Water Quality Control Handbook,* 2000, McGraw Hill

*Manual de Control de la Calidad del Aire,* 2001, Spanish, McGraw-Hill

*Water Quality Control Handbook, Second Edition,* 2007, McGraw-Hill

*Water Quality Control Handbook, Second Edition,* in Chinese, 2015, McGraw-Hill

*A Christian Environmentalist, An Oxymoron or an Obligation?* 2013, Xulon Press

*The Future of Species, The Fantasy of Evolution-The Science of Creation,* 2015, Create Space (Amazon)

*The Mystery of Agape Love,* 2016, Create Space (Amazon)

CPSIA information can be obtained
at www.ICGtesting.com
Printed in the USA
JSHW010350200920
8075JS00003B/6